Trigonometry
DeMYSTiFieD®

DeMYSTiFieD® Series

Accounting Demystified

Advanced Calculus Demystified

Advanced Physics Demystified

Advanced Statistics Demystified

Algebra Demystified

Alternative Energy Demystified

Anatomy Demystified

Astronomy Demystified

Audio Demystified

Biology Demystified

Biophysics Demystified

Biotechnology Demystified

Business Calculus Demystified

Business Math Demystified

Business Statistics Demystified

C++ Demystified

Calculus Demystified

Chemistry Demystified

Chinese Demystified

Circuit Analysis Demystified

College Algebra Demystified

Corporate Finance Demystified

Data Structures Demystified

Databases Demystified

Differential Equations Demystified

Digital Electronics Demystified

Earth Science Demystified

Electricity Demystified

Electronics Demystified

Engineering Statistics Demystified

Environmental Science Demystified

Ethics Demystified

Everyday Math Demystified

Financial Planning Demystified

Forensics Demystified

French Demystified

Genetics Demystified

Geometry Demystified

German Demystified

Home Networking Demystified

Investing Demystified

Italian Demystified

Japanese Demystified

Java Demystified

JavaScript Demystified

Lean Six Sigma Demystified

Linear Algebra Demystified

Logic Demystified

Macroeconomics Demystified

Management Accounting Demystified

Math Proofs Demystified

Math Word Problems Demystified

MATLAB® Demystified

Medical Billing and Coding Demystified

Medical Terminology Demystified

Meteorology Demystified

Microbiology Demystified

Microeconomics Demystified

Minitab Demystified

Nanotechnology Demystified

Nurse Management Demystified

OOP Demystified

Options Demystified

Organic Chemistry Demystified

Personal Computing Demystified

Pharmacology Demystified

Philosophy Demystified

Physics Demystified

Physiology Demystified

Pre-Algebra Demystified

Precalculus Demystified

Probability Demystified

Project Management Demystified

Psychology Demystified

Quality Management Demystified

Quantum Mechanics Demystified

Real Estate Math Demystified

Relativity Demystified

Robotics Demystified

Signals and Systems Demystified

Six Sigma Demystified

Spanish Demystified

Statics and Dynamics Demystified

Statistical Process Control Demystified

Statistics Demystified

Technical Analysis Demystified

Technical Math Demystified

Trigonometry Demystified

Trigonometry
DeMYSTiFieD®

Stan Gibilisco

Second Edition

McGraw Hill

New York   Chicago   San Francisco   Lisbon   London   Madrid   Mexico City
Milan   New Delhi   San Juan   Seoul   Singapore   Sydney   Toronto

McGraw-Hill books are available at special quantity discounts to use as premiums and sales promotions, or for use in corporate training programs. To contact a representative, please e-mail us at bulksales@mcgraw-hill.com.

Trigonometry DeMYSTiFieD®, Second Edition

1 2 3 4 5 6 7 8 9 0   DOC/DOC   1 8 7 6 5 4 3 2

ISBN 978-0-07-178024-7
MHID 0-07-178024-6

Sponsoring Editor
Judy Bass

Editing Supervisor
Stephen M. Smith

Production Supervisor
Richard C. Ruzycka

Acquisitions Coordinator
Bridget L. Thoreson

Project Managers
Nancy Dimitry,
Joanna Pomeranz,
D&P Editorial Services

Composition
D&P Editorial Services

Copy Editor
Joanna Pomeranz,
D&P Editorial Services

Proofreader
Don Dimitry,
D&P Editorial Services

Art Director, Cover
Jeff Weeks

Cover Illustration
Lance Lekander

To Tim, Samuel, and Tony

About the Author

Stan Gibilisco, an electronics engineer, researcher, and mathematician, has authored multiple titles for the McGraw-Hill *Demystified* and *Know-It-All* series, along with numerous other technical books and dozens of magazine articles. His work has been published in several languages.

Contents

		Introduction	*xi*
Part I	**Origins and Theory**		**1**
	CHAPTER 1	**Angles, Distances, and Triangles**	**3**
		Angles and Distances	4
		Triangles	11
		Trigonometric Triangles	17
		Quiz	25
	CHAPTER 2	**Cartesian Coordinates**	**27**
		How It's Assembled	28
		Radial Distance of a Point from the Origin	34
		Distance between Two Points	38
		Finding the Midpoint	43
		Quiz	47
	CHAPTER 3	**The Unit-Circle Paradigm**	**51**
		The Unit Circle	52
		Primary Circular Functions	54
		Secondary Circular Functions	62
		Quiz	72
	CHAPTER 4	**Mappings, Relations, Functions, and Inverses**	**75**
		What's a Mapping?	76
		Types of Mappings	79
		Examples of Relations	83
		Examples of Functions	86

	Inverses of Circular Functions	88
	Graphs of Inverse Circular Functions	93
	Quiz	100
CHAPTER 5	**Hyperbolic Functions**	**103**
	The "Hyper Six"	104
	Hyperbolic Function Graphs	107
	Hyperbolic Inverses	112
	Hyperbolic Inverse Graphs	114
	Quiz	119
CHAPTER 6	**Polar Coordinates**	**121**
	The Variables	122
	Three Basic Graphs	125
	Coordinate Transformations	131
	The Navigator's Way	142
	Quiz	147
CHAPTER 7	**Three-Space and Vectors**	**149**
	Spatial Coordinates	150
	Vectors in the Cartesian Plane	159
	Vectors in the MPC Plane	163
	Vectors in Three Dimensions	168
	Quiz	177
	Test: Part I	**181**
Part II	**Extensions and Applications**	**193**
CHAPTER 8	**Scientific Notation**	**195**
	Formatting Methods	196
	Rules for Use	204
	Approximation, Error, and Precedence	209
	Significant Figures	215
	Quiz	221
CHAPTER 9	**Surveying, Navigation, and Astronomy**	**223**
	Terrestrial Distance Measurement	224
	Interstellar Distance Measurement	229
	Direction Finding and Radiolocation	233
	Quiz	244

CHAPTER 10	**Electrical Waves and Phase**	**247**
	Alternating Current	248
	Phase Angle	252
	Inductive Reactance	258
	Capacitive Reactance	262
	Quiz	266
CHAPTER 11	**Geometrical Optics**	**269**
	Reflection	270
	Refraction	279
	Color Dispersion	287
	Quiz	294
CHAPTER 12	**Trigonometry on a Sphere**	**297**
	The Global Grid	298
	Arcs and Triangles	305
	Global Navigation	316
	Quiz	322
CHAPTER 13	**The Infinite-Series Paradigm**	**325**
	Repeated Addition	326
	Repeated Multiplication	331
	Limits	335
	Expansion of the Sine	338
	Expansion of the Cosine	341
	Expansion of the Tangent	344
	Quiz	349
	Test: Part II	**351**
	Final Exam	*365*
	Answers to Quizzes, Tests, and Final Exam	*391*
	Appendix: Circular and Hyperbolic Identities	*395*
	Suggested Additional Reading	*403*
	Index	*405*

Introduction

This book can help you learn the principles of trigonometry without taking a formal course. It can also serve as a supplemental text in a classroom, tutored, or home-schooling environment. None of the mathematics goes beyond the high-school-senior level (12th grade). If you need a "refresher," you can select from several *Demystified* books dedicated to mathematics topics. In particular, I recommend *Geometry Demystified* as a prerequisite for this course. If you want to build a "rock-solid" mathematics foundation before you start this course, you can go through *Pre-Algebra Demystified*, *Algebra Demystified*, *Geometry Demystified*, and *Algebra Know-It-All*. If you want to extend your knowledge beyond the scope of this course, I recommend *Pre-Calculus Know-It-All* and *Calculus Know-It-All*.

How to Use This Book

This book contains abundant multiple-choice questions written in standardized test format. You'll find an "open-book" quiz at the end of every chapter. You may (and should) refer to the chapter texts when taking these quizzes. Write down your answers, and then give your list of answers to a friend. Have your friend reveal your score, but not which questions you missed. The correct answer choices appear in the back of the book. Stick with a chapter until you get most of the quiz answers correct.

Two major parts constitute this course. Each part ends with a multiple-choice test. Take these tests when you're done with the respective parts and have taken all the chapter quizzes. Don't look back at the text when taking the part tests. They're easier than the chapter-ending quizzes, and they don't require you to memorize trivial things. A satisfactory score is three-quarters correct. The correct answer choices appear in the back of the book.

The course concludes with a 100-question final exam. Take it when you've finished all the parts, all the part tests, and all of the chapter quizzes. A satisfactory score is at least 75 percent correct answers. Again, the correct answer choices are listed in the back of the book.

With the part tests and the final exam, as with the quizzes, have a friend divulge your score without letting you know which questions you missed. That way, you won't subconsciously memorize the answers. You might want to take each test, and the final exam, two or three times. When you get a score that makes you happy, you can (and should) check to see where your strengths and weaknesses lie.

I've posted explanations for the chapter-quiz answers (but not for the part-test or final-exam answers) on the Internet. As we all know, Internet particulars change; but if you conduct a phrase search on "Stan Gibilisco," you'll get my Web site as one of the first hits. You'll find a link to the explanations on that site. As of this writing, it's **www.sciencewriter.net**.

Strive to complete one chapter of this book every 10 days or two weeks. Don't rush, but don't go too slowly either. Proceed at a steady pace and keep it up. That way, you'll complete the course in a few months. (As much as we all wish otherwise, nothing can substitute for "good study habits.") When you've finished up, you can use this book as a permanent reference.

I welcome your ideas and suggestions for future editions.

Stan Gibilisco

Trigonometry
DeMYSTiFieD®

Part I

Origins and Theory

chapter **1**

Angles, Distances, and Triangles

The word *trigonometry* comes from the Sanskrit expression "triangle measurement." In its simplest form, trigonometry involves the relationships between the sides and angles of triangles where one of the angles measures 90° (a *right angle*). But, as you'll learn as you go through this book, trigonometry involves a lot more than triangles! Before we get into "real trigonometry," let's review the basic principles and jargon for angles, distances, and triangles on *Euclidean* (flat) surfaces.

CHAPTER OBJECTIVES

In this chapter, you will

- Quantify angles in radians and degrees.
- Learn how lines, angles, and distances relate on flat surfaces.
- Classify triangles according to their interior angles.
- Define the trigonometric functions as ratios among lengths of the sides of a right triangle.
- Learn the Pythagorean theorem for right triangles.

Angles and Distances

When two lines intersect, we get four distinct angles at the point of intersection. In most cases, we'll find that two of the angles are "sharp" and two are "dull." If all four of the angles happen to be identical, then they all constitute right angles, and we say that the lines run *perpendicular, orthogonal,* or *normal* to each other at the intersection point. We can also define an angle using three points connected by two line segments. In that case, the angle appears between the line segments.

Naming Angles

The italic, lowercase Greek letter *theta* has become popular as an "angle name." It looks like an italic numeral *0* with a horizontal line through it (θ). When writing about two different angles, a second Greek letter can go along with θ. The italic, lowercase letter *phi* works well for this purpose. It looks like an italic lowercase English letter *o* with a forward slash through it (ϕ).

Sometimes, mathematicians use the italic, lowercase Greek letters *alpha*, *beta,* and *gamma* to represent angles. We write these symbols as α, β, and γ, respectively. When things get messy and we have a lot of angles to write about, we can add numeric subscripts to italic Greek letters and denote our angles as symbols such as θ_1, θ_2, θ_3, and so on, or α_1, α_2, α_3, and so on.

TIP   *If you don't like Greek symbols, you can represent angle variables with more familiar characters such as x, y, and z. As long as you know the context and stay consistent in a given situation, you can call an angle anything you want.*

Radian Measure

Imagine two *rays* (or "half-lines") pointing out from the center of a circle so that each ray intersects the circle at a specific point. Suppose that the distance between these points, as measured along the curve of the circle (not along a straight line), is the same as the circle's radius. In that case, the angle between the rays measures one *radian* (1 rad).

A circle has 2π rad going exactly once around, where π (the lowercase Greek letter pi, not in italics) stands for the ratio of a circle's circumference to its diameter. This number is the same for all perfect circles in a flat plane. The number π is *irrational*, meaning that we can't express it as a ratio between whole numbers. It equals approximately 3.14159.

Mathematicians commonly use the radian as their standard unit of angular measure. Sometimes, they'll omit the "rad" after an angle when they know that they're working with radians (and when they're sure that you know it too). Based on that convention, we can sum up the nature of the radian as follows:

- An angle of $\pi/2$ represents 1/4 of a circle
- An angle of π represents 1/2 of a circle
- An angle of $3\pi/2$ represents 3/4 of a circle
- An angle of 2π represents a full circle

We can categorize angles into several types:

- An *acute angle* has a measure of more than 0 but less than $\pi/2$
- A *right angle* has a measure of exactly $\pi/2$
- An *obtuse angle* has a measure of more than $\pi/2$ but less than π
- A *straight angle* has a measure of exactly π
- A *reflex angle* has a measure of more than π but less than 2π

Degree Measure

The *angular degree* (°), also called the *degree of arc*, is the unit of angular measure familiar to lay people. One degree (1°) represents 1/360 of a full circle. We can summarize degree measure by noting a few facts:

- An angle of 90° represents 1/4 of a circle
- An angle of 180° represents 1/2 of a circle
- An angle of 270° represents 3/4 of a circle
- An angle of 360° represents a full circle

When we use degrees, the general angle types break down as follows:

- An acute angle has a measure of more than 0 but less than 90°
- A right angle has a measure of exactly 90°
- An obtuse angle has a measure of more than 90° but less than 180°
- A straight angle has a measure of exactly 180°
- A reflex angle has a measure of more than 180° but less than 360°

TIP *Whenever you see a quantitative (numerical) reference to an angle and no unit goes with it, and especially if you see the symbol π in the expression, you can assume that the author wants to express the angle in radians. However, you should always check the context to make certain! If you're writing a paper and you want to express* *an angle in radians, you can write "rad" after the quantity for the angle value. That way, you can make sure that your readers won't get confused. If you write "2 rad," for example, your readers will know that you mean two radians, not two degrees. Similarly, if you want your reader to know for sure that you mean to express an angle in degrees, you should put a little degree symbol after the quantity.*

Minutes and Seconds of Arc

Once in awhile, you'll read about fractions of a degree called *minutes of arc* or *arc minutes*. One arc minute (symbolized as 1 arc min or 1′) equals 1/60 of a degree. You might also read about *seconds of arc*, also known as *arc seconds*. One arc second (symbolized as 1 arc sec or 1″) equals 1/60 of an arc minute, or 1/3600 of a degree.

TIP   *Arc minutes and arc seconds differ from the minutes and seconds that astronomers sometimes use for defining the positions of objects in the sky. You'll learn about those minutes and seconds in Chap. 7.*

 **PROBLEM 1-1**

Imagine that the measure of a certain angle θ equals $\pi/6$. What fraction of a complete circular rotation does this angle represent? What's the measure of θ in degrees?

 **SOLUTION**

A full circular rotation represents an angle of 2π. The value $\pi/6$ equals 1/12 of 2π. Therefore, the angle θ represents 1/12 of a full circle. In degree measure, that's 1/12 of 360°, or 30°.

 Still Struggling

If you've always measured and defined angles in degrees, the radian can seem strange at first. "Why," you ask, "would anyone want to divide a circle into an *irrational* number of angular parts?" Mathematicians prefer the radian-measure system because it works out more simply (believe it or not) than the degree-measure scheme in some disciplines. The radian is actually a *more natural* angular quantity than the degree! We can define the radian in geometric terms without talking about any numbers whatsoever, just as we can define the diagonal of a square as the distance from one corner to the opposite corner. The radian is a purely geometric unit. The degree was contrived by humans, probably dating back to ancient times when people knew that a year contained roughly 360 days. When we think of a year as a "circle in time," it's easy to extend the idea to all geometric circles. If our distant ancestors had seriously considered the true length of the year, which comes out to a fractional number of days (and, one might argue, an irrational number that doesn't even stay the same as the centuries pass), maybe they'd have chosen some other standard angle to represent the degree, such as 1/100 of a circle or 1/1000 of a circle. Even so, the radian is the most natural unit possible!

Angle Notation

Imagine that P, Q, and R represent three distinct points. Let L represent the line segment connecting P and Q, and let M represent the line segment connecting R and Q. We can denote the angle between L and M, as measured at point Q in the plane defined by the three points, by writing $\angle PQR$ or $\angle RQP$ as shown in Fig. 1-1.

If we want to specify the *rotational sense* of the angle, then $\angle RQP$ indicates the angle as we turn counterclockwise from M to L, and $\angle PQR$ indicates the angle as we turn clockwise from L to M. We consider counterclockwise-going angles as having positive values, and clockwise-going angles as having negative values.

In the situation of Fig. 1-1, $\angle RQP$ is positive while $\angle PQR$ is negative. If we make an approximate guess as to the measures of the angles in Fig. 1-1, we might say that $\angle RQP = +60°$ while $\angle PQR = -60°$.

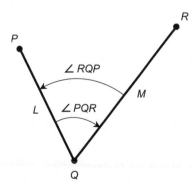

FIGURE 1-1 · Angle notation.

TIP *Rotational sense doesn't make any difference in basic geometry. However, it does matter when we work in* coordinate geometry *involving graphs. We'll get into coordinate geometry, also known as* analytic geometry, *later in this book. For now, let's not worry about the rotational sense in which we express or measure an angle. We can consider all angles as having positive measures.*

Angle Bisection Principle

Let's look at an angle called $\angle PQR$ that measures less than 180°, and that we can define with three points P, Q, and R as shown in Fig. 1-2. In this situation, there's *exactly one* (in other words, *one and only one*) ray M that *bisects* $\angle PQR$ (divides $\angle PQR$ in half). If S represents a point on M other than point Q, then $\angle PQS = \angle SQR$.

Perpendicular Principle

Consider a line L that passes through two distinct points P and Q as shown in Fig. 1-3. Suppose that R represents a point that doesn't lie on L. In that case, there's exactly one line M through point R, intersecting line L at some point S, such that M runs perpendicular to L (M and L intersect at a right angle) at point S.

Perpendicular Bisector Principle

When we have a line segment connecting two points P and R, we can always find exactly one line M that runs perpendicular to line segment PR and that

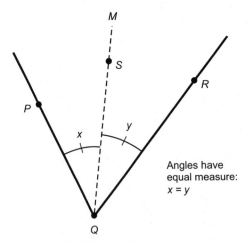

FIGURE 1-2 • The angle bisection principle.

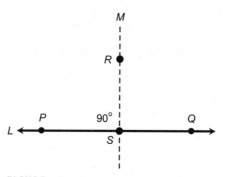

FIGURE 1-3 • The perpendicular principle.

intersects PR at a point Q, such that the distance from P to Q (which we can write as PQ) equals the distance from Q to R (which we denote as QR). In other words, every line segment has exactly one *perpendicular bisector*. Figure 1-4 illustrates this principle.

Distance Addition and Subtraction

Imagine that P, Q, and R represent points on a line, arranged so that Q lies somewhere between P and R. The following equations hold true concerning the distances between pairs of points as we measure them along the line as shown in Fig. 1-5:

$$PQ + QR = PR$$
$$PR - PQ = QR$$
$$PR - QR = PQ$$

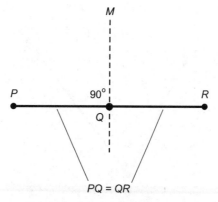

FIGURE 1-4 • The perpendicular bisector principle. Illustration for Problem 1-2.

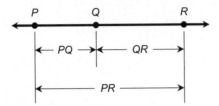

FIGURE 1-5 · Distance addition and subtraction.

Angular Addition and Subtraction

Suppose that P, Q, R, and S represent points that all lie in the same *Euclidean plane*. In other words, all four points lie on a single, perfectly flat surface. Let Q represent the convergence point of three angles $\angle PQR$, $\angle PQS$, and $\angle SQR$, with line segment QS between line segments QP and QR as shown in Fig. 1-6. In that case, we'll always find that the following equations hold true:

$$\angle PQS + \angle SQR = \angle PQR$$

$$\angle PQR - \angle PQS = \angle SQR$$

$$\angle PQR - \angle SQR = \angle PQS$$

PROBLEM 1-2

Go back to Fig. 1-4 and examine it again. Imagine some point S, other than point Q, that lies on line M (the perpendicular bisector of line segment PR). What can you say about the lengths of line segments PS and SR?

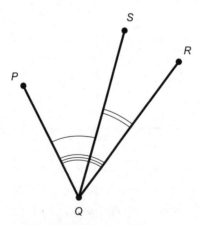

FIGURE 1-6 · Angular addition and subtraction. Illustration for Problem 1-3.

 **SOLUTION**

You can "streamline" the solutions to problems like this by making your own drawings. With the aid of your own sketch, you should see that for any point *S* that lies on line *M* (other than point *Q*), the distance *PS* exceeds the distance *PQ* (that is, *PS* > *PQ*), and the distance *SR* exceeds the distance *QR* (that is, *SR* > *QR*).

 PROBLEM 1-3

Look again at Fig. 1-6. Suppose that you move point *S* either straight toward yourself (out of the page) or straight away from yourself (back behind the page). In either case, point S no longer lies in the same plane as points *P, Q,* and *R* do. What can you say about the measures of ∠*PQR*, ∠*PQS*, and ∠*SQR*?

 **SOLUTION**

You can use your "three-dimensional mind's eye" to envision these situations. Either way, you should be able to see that the sum of the measures of ∠*PQS* and ∠*SQR* exceeds the measure of ∠*PQR* because the measures of ∠*PQS* and ∠*SQR* both increase if point *S* departs *at a right angle* from the plane containing points *P, Q,* and *R.*

Triangles

In technical terms, a *triangle* consists of three line segments, joined pairwise at their end points, and including those end points. In order to determine a triangle, the three points must not be *collinear* (they must not all lie on the same straight line). For now, let's assume that the surface or *universe* for all our triangles is Euclidean ("flat"), and not "curved" like the surface of a sphere, cone, or cylinder.

TIP   *In a Euclidean universe, we can always determine the shortest distance between two points by finding the straight line segment connecting the points, and then measuring the length of the line segment. Conversely, if the shortest distance between two points always constitutes a straight line, then we know that we're working in a Euclidean universe.*

Vertices, Names, and Sides

Figure 1-7 shows three points called *A*, *B*, and *C*, connected by line segments to form a triangle. We call each point a *vertex* of the triangle, so the figure has three *vertices* (or *vertexes*). We call the whole figure "triangle *ABC*."

When naming triangles, geometers write an uppercase Greek letter delta (Δ) in place of the word "triangle." According to that notation, Fig. 1-7 portrays Δ*ABC*. Alternatively, we can call it Δ*BCA*, Δ*CAB*, Δ*CBA*, Δ*BAC*, or Δ*ACB*.

We can name the sides of the triangle in Fig. 1-7 according to their end points. In this case, our triangle has three sides: line segment *AB* (or *BA*), line segment *BC* (or *CB*), and line segment *CA* (or *AC*).

Interior Angles

Each vertex of a triangle corresponds to a specific *interior angle*, which always measures more than 0° (or 0 if we express it in radians) but less than 180° (or π if we express it in radians). In Fig. 1-7, we denote the interior angles using the lowercase italic English letters *x*, *y*, and *z*. Alternatively, we can use italic lowercase Greek letters to symbolize the interior angles. Subscripts can help us distinguish the angles from one another; for example, θ_a, θ_b, and θ_c could represent the interior angles at vertices *A*, *B*, and *C*, respectively.

Directly Similar Triangles

Two triangles are *directly similar* if and only if they have the same proportions in the same rotational sense (that is, as we go around them both in the same direction). Two triangles *are not* directly similar if we must flip one of them over, in addition to changing its size and rotating it, in order to place it exactly over the other one.

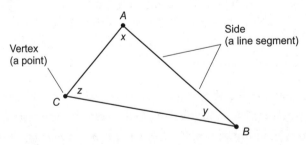

FIGURE 1-7 · Vertices, sides, and angles of a triangle.

Directly Congruent Triangles

Two triangles are *directly congruent* if and only if they're directly similar and the corresponding sides have identical lengths. If we take one of the triangles and rotate it clockwise or counterclockwise to the correct extent, we can "paste" it precisely over the other one. Rotation and motion are allowed, but flipping-over, also called *mirroring*, is forbidden.

Still Struggling

You should remember these fundamental facts when you work with triangles that look a lot alike:

- If two or more triangles are directly congruent, then the corresponding sides have equal lengths as you proceed around them all in the same direction. The converse also holds true. If two or more triangles have corresponding sides with equal lengths as you proceed around them all in the same direction, then all the triangles are directly congruent.
- If two or more triangles are directly congruent, then the corresponding interior angles have equal measures as you proceed around them all in the same direction. However, the converse does not always hold true. Two or more triangles can have corresponding interior angles with equal measures when you go around them all in the same direction, and nevertheless fail to be directly congruent (although they'll always be directly similar).

Acute Triangle

We have an *acute triangle* if and only if each of the three interior angles is acute. In such a triangle, none of the angles measure as much as a right angle (90° or π/2); they're all smaller than that. Figure 1-8 shows some examples.

Obtuse Triangle

We have an *obtuse triangle* if and only if one of the three interior angles is obtuse, measuring more than a right angle (90° or π/2) but less than a straight angle (180° or π). In a triangle of this type, the two nonobtuse angles are both acute. Figure 1-9 shows some examples.

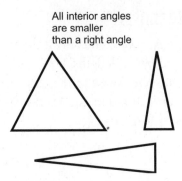

All interior angles
are smaller
than a right angle

FIGURE 1-8 • In an acute triangle, all angles measure less than a right angle.

Isosceles Triangle

Imagine a triangle with sides that have lengths s, t, and u. Let x represent the angle opposite the side of length s, let y represent the angle opposite the side of length t, and let z represent the angle opposite the side of length u. Now suppose that *at least one* of the following equations holds true:

$$s = t$$
$$t = u$$
$$s = u$$
$$x = y$$
$$y = z$$
$$x = z$$

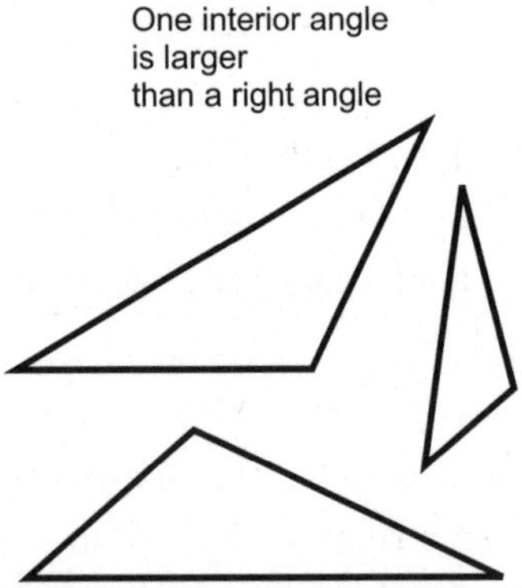

One interior angle
is larger
than a right angle

FIGURE 1-9 • In an obtuse triangle, one angle measures more than a right angle.

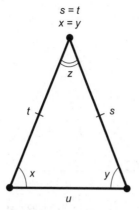

FIGURE 1-10 · An isosceles triangle.

Figure 1-10 shows an example of such a situation, where $s = t$. Whenever we find a triangle that has two sides of identical length, we call it an *isosceles triangle*. For sides and angles in the orientation of Fig. 1-10:

$$s = t \leftrightarrow x = y$$

Still Struggling

In a logical statement, a double-headed arrow ($\leftrightarrow$) stands for the expression "if and only if," and a single-headed arrow pointing to the right ($\rightarrow$) stands for the expression "logically implies." For example, when we write

$$s = t \leftrightarrow x = y$$

we assert that

$$s = t \rightarrow x = y$$

and also that

$$x = y \rightarrow s = t$$

Mathematicians sometimes abbreviate "if and only if" as "iff," meaning that the logical implication works both ways. In mathematics and logic, when we claim that "A implies B," we mean "If A holds true, then B always holds true," or "If A, then B."

Equilateral Triangle

Imagine a triangle with sides of lengths s, t, and u. Let x represent the angle opposite the side of length s, let y represent the angle opposite the side of length

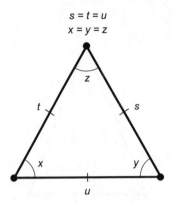

FIGURE 1-11 · An equilateral triangle.

t, and let z represent the angle opposite the side of length u. Suppose that either of the following equations holds true:

$$s = t = u$$

or

$$x = y = z$$

In this case we have an *equilateral triangle* (Fig. 1-11), and we can make the logical statement

$$s = t = u \leftrightarrow x = y = z$$

TIP   *Any two equilateral triangles chosen "at random" have identical shape, even if their sizes differ. All equilateral triangles are directly similar to each other.*

Right Triangle

Imagine a triangle $\triangle PQR$ with sides having lengths s, t, and u as shown in Fig. 1-12. We call this figure a *right triangle* if and only if one of its interior angles

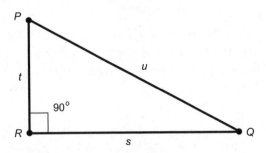

FIGURE 1-12 · A right triangle.

constitutes a right angle (90° or $\pi/2$). Figure 1-12 illustrates a right triangle in which $\angle QRP$ forms the right angle. The side opposite the right angle has the longest length (in this case u); we call it the *hypotenuse* of the right triangle.

Sum of Angle Measures

In any triangle on a Euclidean surface, the sum of the measures of the interior angles equals 180° (or π, if we express it in radians). This principle holds true whether the figure is a right triangle or not.

Trigonometric Triangles

If we want to quantify all the dimensions of an acute or obtuse triangle, we can break it into two right triangles by "dropping a perpendicular" from the vertex having the largest angle to the side opposite that angle. Then we can deduce the lengths of all the sides and the measures of all the angles—if we have a certain minimum amount of information. For the rest of this chapter, let's confine our attention to right triangles.

Ratios of Sides

Consider a right triangle defined by three vertex points P, Q, and R, along with a specific angle θ as shown in Fig. 1-13. Suppose that the angle at vertex R is the right angle. Let s represent the length of line segment QR (the base in this case), let t represent the length of line segment RP (the height in this case), and let u represent the length of line segment PQ (the hypotenuse in this case). Let θ represent $\angle RQP$, the angle between line segments QR and QP. We define the *sine* (sin), *cosine* (cos), *tangent* (tan), *cosecant* (csc), *secant* (sec), and *cotangent* (cot), commonly called *trigonometric functions*, as follows:

$$\sin \theta = t/u$$

$$\cos \theta = s/u$$

$$\tan \theta = t/s$$

$$\csc \theta = u/t$$
$$= 1/(\sin \theta)$$

$$\sec \theta = u/s$$
$$= 1/(\cos \theta)$$

$$\cot \theta = s/t$$
$$= 1/(\tan \theta)$$

In plain language, we can state the above definitions as follows:

- The sine of θ equals the height divided by the hypotenuse length.
- The cosine of θ equals the base length divided by the hypotenuse length.
- The tangent of θ equals the height divided by the base length.
- The cosecant of θ equals the hypotenuse length divided by the height; it's the reciprocal of the sine.
- The secant of θ equals the hypotenuse length divided by the base length; it's the reciprocal of the cosine.
- The cotangent of θ equals the base length divided by the height; it's the reciprocal of the tangent.

The Theorem of Pythagoras

One of the most famous mathematical facts ever discovered is the *Theorem of Pythagoras*, also called the *Pythagorean theorem* (named after *Pythagoras of Samos*, a Greek mathematician who lived in the sixth century B.C.). Some debate goes on as to whether Pythagoras of Samos really proved this theorem before anyone else did, but various proofs exist today.

Imagine a right triangle defined by points P, Q, and R whose sides have lengths s, t, and u respectively as shown in Fig. 1-13. Let u represent the length of the hypotenuse. In this situation, the Theorem of Pythagoras tells us that

$$u^2 = s^2 + t^2$$

Conversely, if we find a triangle on a flat surface such that the sum of the squares of the lengths of any two sides equals the square of the length of the remaining side, then that figure will always turn out to be a right triangle.

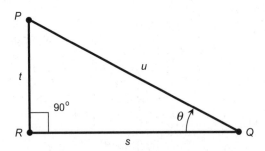

FIGURE 1-13 · We can define the trigonometric functions for an angle θ within a right triangle using this diagram. Illustration for Problems 1-4, 1-5, and 1-6.

Proving It

Imagine four directly congruent right triangles, each of which has sides of lengths *s*, *t*, and *u* units, with the hypotenuses all measuring *u* units. Let's arrange them as shown in Fig. 1-14 so they define a large square (heavy lines) whose sides each measure *s* + *t* units long. As we learned in basic geometry, we can calculate the area of the large square region by squaring the length of one side. If we call the area of the large square A_L, then

$$A_L = (s + t)^2$$

which multiplies out to

$$A_L = s^2 + 2st + t^2$$

where we express A_L in *square units* or *units squared*.

The unshaded region inside the large square forms a smaller square measuring *u* units on each side. We can calculate its area A_U as

$$A_U = u^2$$

Still Struggling

"How," you ask, "do we know that the unshaded region in Fig. 1-14 actually constitutes a square?" That's an excellent question! Using the knowledge that you gained in first-year geometry, you can prove that fact by showing that all four of the unshaded region's interior angles measure 90°. Then, because all the sides have the same length *u*, you must conclude that the figure forms a square. Try this proof as an "extra credit" exercise. Here's a hint: Take advantage of the fact that the sum of the measures of the interior angles of a triangle always equals 180°. Here's another hint: Once you've proven that any one of the interior angles of the unshaded region measures 90°, you can prove that each of the other three interior angles also measures 90° by repeating the first proof verbatim three times!

Demonstrating the Theorem of Pythagoras (*continued*)

Each of the shaded right triangles in Fig. 1-14 has an area A_T that equals half the height times the length of the base. (That's another thing that we learned

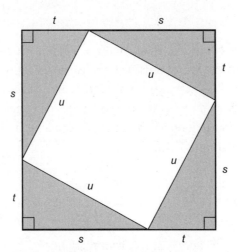

FIGURE 1-14 · Proof of the Theorem of Pythagoras. We can arrange four directly congruent right triangles to form a square within a square.

in basic geometry.) In the situation of Fig. 1-14, if we call s the base length and t the height for each shaded right triangle, then

$$A_T = st/2$$

We have four of these triangles, all congruent. Let's call the sum of their areas A_S. Then we know that

$$A_S = 4A_T$$
$$= 4\,(st/2)$$
$$= 2st$$

We can see that the sum of the shaded and unshaded regions equals the area of the large square, so that

$$A_S + A_U = A_L$$

Let's make three substitutions in the above equation:

- Replace A_S with its equivalent quantity $2st$
- Replace A_U with its equivalent quantity u^2
- Replace A_L with its equivalent quantity $s^2 + 2st + t^2$

These changes give us the equation

$$2st + u^2 = s^2 + 2st + t^2$$

We can subtract the quantity $2st$ from the expressions on each side of the equals sign to get the Pythagorean theorem formula

$$u^2 = s^2 + t^2$$

Still Struggling

If you want to avoid symbology, you can state the Pythagorean theorem as follows: "The square of the length of the hypotenuse of any right triangle equals the sum of the squares of the lengths of the other two sides." This fact holds true, however, only in *Euclidean geometry*, when the triangle lies on a flat surface. It does not hold for triangles on, say, the surface of a sphere or cone. We're dealing only with Euclidean geometry right now. We won't concern ourselves with other geometries—yet!

Range of Angles

In the system we've worked with so far, we can define the values of "trig functions" only for angles between (but not always including) $0°$ and $90°$ (0 and $\pi/2$). All angles outside this range are better dealt with using a more comprehensive methodology, which we'll examine in Chap. 3. The angle limitation represents the main shortcoming of the so-called *right-triangle paradigm* for trigonometric functions.

Using the right-triangle scheme, we can't define a trigonometric function if the denominator equals 0 in its side-length ratio. The length of the hypotenuse is always nonzero and positive, but if we "squash" or "squeeze" a right triangle flat either horizontally or vertically, then the length of one of the *adjacent sides* (of length s or t as shown in Fig. 1-13) can equal 0. Such objects aren't true triangles, but some people like to include them to account for angles of $0°$ and $90°$ (or 0 and $\pi/2$).

 PROBLEM 1-4

Imagine a triangle whose sides measure exactly 3, 4, and 5 units long. What's the sine of the angle θ opposite the 3-unit side?

 SOLUTION

Before we do any calculations, let's make sure that a triangle with sides of 3, 4, and 5 units actually constitutes a right triangle! We can test it by checking to see if the Pythagorean theorem applies. If this triangle is a right triangle, then the side measuring 5 units will form the hypotenuse, so we should find that

$$3^2 + 4^2 = 5^2$$

Checking, we see that $3^2 = 9$ and $4^2 = 16$. Therefore

$$3^2 + 4^2 = 9 + 16$$
$$= 25$$
$$= 5^2$$

It's a right triangle, sure enough! It helps to draw a picture here, after the fashion of Fig. 1-13 on page 18. We place the angle θ, which we want to analyze, at vertex Q. Then we label the hypotenuse $u = 5$, and we label the other two sides $s = 4$ and $t = 3$. According to the formulas above, we have

$$\sin \theta = t/u$$
$$= 3/5$$

PROBLEM 1-5

What are the values of the other five circular functions of θ as defined in Problem 1-4?

 SOLUTION

Now that we know the relative orientations of the angle θ and the sides s, t, and u (according to Fig. 1-13), we can input numbers to the side-ratio formulas to obtain the following values:

$$\cos \theta = s/u$$
$$= 4/5$$
$$\tan \theta = t/s$$
$$= 3/4$$
$$\csc \theta = u/t$$
$$= 5/3$$
$$\sec \theta = u/s$$
$$= 5/4$$
$$\cot \theta = s/t$$
$$= 4/3$$

Pythagorean Theorem for Sine and Cosine

The sum of the squares of the sine and cosine of an angle θ always equals 1, as long as we can construct a right triangle that includes θ as one of the acute angles. We can write this fact as the equation

$$\sin^2 \theta + \cos^2 \theta = 1$$

In this context, the expression $\sin^2 \theta$ means that we take the sine of the angle and then square the result. By convention,

$$\sin^2 \theta = (\sin \theta)^2$$

The same notation applies to the cosine, tangent, cosecant, secant, cotangent, and for all other similar expressions that you'll see in this book.

 PROBLEM 1-6

Derive the foregoing formula for the sine and cosine from the definitions of those functions (as ratios of sides of a right triangle) and the original Pythagorean theorem.

 SOLUTION

Look once again at Fig. 1-13 on page 18, and observe the relative orientations of s, t, u, θ, and the right angle. The ratio formulas for the trigonometric functions tell us that

$$\sin \theta = t / u$$

and

$$\cos \theta = s / u$$

When we square both of these quantities, we get

$$\sin^2 \theta = (t / u)^2$$
$$= t^2 / u^2$$

and

$$\cos^2 \theta = (s / u)^2$$
$$= s^2 / u^2$$

When we add the square of the sine to the square of the cosine, we have

$$\sin^2 \theta + \cos^2 \theta = t^2/u^2 + s^2/u^2$$
$$= (t^2 + s^2)/u^2$$
$$= (s^2 + t^2)/u^2$$

According to the theorem, we know that

$$s^2 + t^2 = u^2$$

We can now make a final substitution to get

$$\sin^2 \theta + \cos^2 \theta = u^2/u^2$$
$$= 1$$

QUIZ

Refer to the text in this chapter if necessary. A good score is eight correct. Answers are in the back of the book.

1. **Based on the definitions of direct similarity and direct congruence given in this chapter, which of the following statements holds true in general?**
   A. If two triangles are directly similar, then they are directly congruent.
   B. If two triangles are directly congruent, then they are directly similar.
   C. Two triangles are directly congruent if and only if they are directly similar.
   D. None of the above

2. **Based on the definition given in this chapter, a radian measures approximately**
   A. $57.3°$.
   B. $114.6°$.
   C. $28.6°$.
   D. $70.7°$.

3. **Figure 1-15 illustrates a triangle that we call $\triangle ABC$, with angles having measures as shown. Based on this information, we can have complete confidence that $\triangle ABC$ constitutes**
   A. an acute triangle.
   B. a right triangle.
   C. an obtuse triangle.
   D. a reflex triangle.

4. **In the situation of Fig. 1-15, we can be absolutely certain that**
   A. $x < 90°$.
   B. $y < 90°$.
   C. $x + y = 90°$.
   D. All of the above

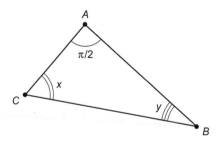

FIGURE 1-15 • Illustration for Quiz Questions 3 and 4.

5. Which of the following statements holds true in general?
   A. All right triangles are equilateral.
   B. All isosceles triangles are equilateral.
   C. All equilateral triangles are isosceles.
   D. All obtuse triangles are isosceles.

6. Consider an angle θ inside a right triangle, where θ is not the right angle. Based on the definitions of the trigonometric functions given in this chapter, which of the following inequalities holds true in general?
   A. $0 < \sin \theta < 1$
   B. $\sin \theta = 1$
   C. $\sin \theta > 1$
   D. $\sin \theta = 0$

7. Consider an angle θ inside a right triangle, where θ is not the right angle. Based on the definitions of the trigonometric functions given in this chapter, which of the following inequalities holds true in general?
   A. $0 < \cos \theta < 1$
   B. $\cos \theta = 1$
   C. $\cos \theta > 1$
   D. $\cos \theta = 0$

8. Consider an angle θ inside a right triangle, where θ is not the right angle. Based on the definitions of the trigonometric functions given in this chapter, which of the following inequalities holds true in general?
   A. $0 < \tan \theta < 1$
   B. $\tan \theta < 1$
   C. $\tan \theta > 1$
   D. $\tan \theta > 0$

9. Consider an angle θ inside a right triangle, where θ is not the right angle. Based on the definitions of the trigonometric functions given in this chapter, which of the following inequalities holds true in general?
   A. $0 < \csc \theta < 1$
   B. $\csc \theta < 1$
   C. $\csc \theta > 1$
   D. $\csc \theta > 0$

10. Imagine that the measure of a certain angle θ equals $\pi/8$ rad. What fraction of a complete circular rotation does this angle represent?
    A. 1/4
    B. 1/8
    C. 1/16
    D. 1/24

chapter 2

Cartesian Coordinates

If you've taken a course in algebra or geometry, you've learned about the graphing system called *Cartesian* (pronounced "car-TEE-zhun") *two-space*, also known as *Cartesian coordinates* or the *Cartesian plane*. This system's name derives from its discoverer, the French mathematician *Rene Descartes* (1596–1650).

CHAPTER OBJECTIVES

In this chapter, you will

- Set up a system of Cartesian coordinates using two number lines.
- Contrast the independent-variable and dependent-variable axes.
- Distinguish functions from relations by examining their graphs.
- Calculate the distances of points from the coordinate origin.
- Calculate the distances between pairs of points.
- Locate the midpoints of line segments connecting pairs of points.

How It's Assembled

We can put together a Cartesian plane by positioning two identical *real-number lines* so that they run perpendicular to each other and intersect at their zero points. The lines cross at the *coordinate origin* (or simply the *origin*). Each number line forms an *axis* that can represent the values of a mathematical variable.

The Variables

Figure 2-1 shows a simple Cartesian plane. We portray one variable along a horizontal line and the other variable along a vertical line. In a true Cartesian system of coordinates, we must graduate the number-line scales in increments of the same size. Here, the horizontal axis represents the *independent variable* or "input," and the vertical axis represents the *dependent variable* or "output."

Figure 2-2 shows the same coordinate plane as Fig. 2-1 does, but now we've labeled the axes with variable names. We call the independent variable x and the dependent variable y. Each point on the plane corresponds to a unique *ordered pair* of variable values, and each ordered pair of variable values corresponds to a unique point on the plane. In Fig. 2-2, we've plotted several ordered pairs of the form (x,y).

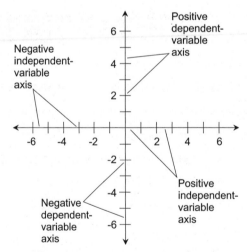

FIGURE 2-1 • The Cartesian plane consists of two real-number lines intersecting at a right angle, forming axes for the variables.

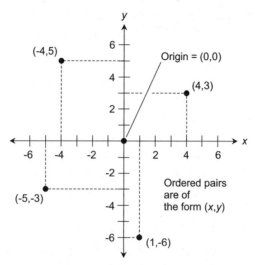

FIGURE 2-2 · Five ordered pairs (including the origin) plotted as points on the Cartesian plane. The dashed lines clarify the axis locations.

TIP   *The term "ordered" refers to the fact that it makes a difference which value we list first and which value we list second. We always separate the two values with a comma. We don't leave a space after the comma (as we would do in an ordinary list).*

Still Struggling

In a Cartesian plane, the values in an ordered pair represent the *coordinates* of a point on the plane. People sometimes talk as if the ordered pair *is* the point, but in strict terms the ordered pair simply *names* the point. We call the independent-variable value the *abscissa*. We call the dependent-variable value the *ordinate*.

Interval Notation

In precalculus and calculus, we'll often want to express a continuous span of values that a variable can attain. We call such a span an *interval*. An interval always has a certain minimum value and a certain maximum value representing the *extremes*. Consider the following four expressions:

$$0 < x < 2$$
$$-1 \le y < 0$$
$$4 < z \le 8$$
$$-\pi \le \theta \le \pi$$

These expressions have the following meanings, in order:

- The value of x is larger than 0 but smaller than 2.
- The value of y is larger than or equal to −1 but smaller than 0.
- The value of z is larger than 4 but smaller than or equal to 8.
- The value of θ is larger than or equal to −π but smaller than or equal to π.

The first case gives us an example of an *open interval*, which we can write as

$$x \in (0,2)$$

which translates to "x is an element of the open interval (0,2)." Don't mistake this open interval for an ordered pair! The notations look the same, but the meanings are completely different.

The second and third cases provide us with examples of *half-open intervals*. We denote this type of interval with a square bracket on the side of the *included value* and a rounded parenthesis on the side of the *excluded value*. For the second expression we can write

$$y \in [-1,0)$$

which means "y is an element of the half-open interval [−1,0)," and for the third expression we can write

$$z \in (4,8]$$

which means "z is an element of the half-open interval (4,8]."

The fourth case constitutes an example of a *closed interval*. We use square brackets on both sides to show that both extremes are included. We can denote this interval by writing

$$\theta \in [-\pi,\pi]$$

which translates to "θ is an element of the closed interval [−π,π]."

Relations and Functions

Do you remember the definitions of the terms *relation* and *function* from your algebra courses? We can define a relation as an action or process that transforms (or *maps*) values of some variable to values of another variable. We can define a function as a relation in which we never have more than one value of the dependent variable for any single value of the independent variable. In other words, there can exist *at most* one "output" for any "input." (If a particular "input" value produces no "output," that's okay.)

TIP *The Cartesian plane provides an excellent way to illustrate relations and functions. Later in this course we'll take a closer look at how relations and functions arise in mathematics, and how we can "twist" them so that they work "backward," a trick that enjoys special significance in trigonometry.*

 Still Struggling

In a true Cartesian plane, both axes are *linear*, and both axes are graduated in *increments* of the same size. The term *linear axis* refers to the fact that the change in value is always directly proportional to the physical displacement along the axis. For example, if we travel 5 millimeters along an axis and the value changes by 1 unit, then that fact holds true everywhere along that axis.

The Quadrants

Any pair of intersecting lines divides a plane into four parts. In the Cartesian system, we call these parts *quadrants* as shown in Fig. 2-3.

- In the *first quadrant*, both variables are positive.
- In the *second quadrant*, the independent variable is negative and the dependent variable is positive.
- In the *third quadrant*, both variables are negative.
- In the *fourth quadrant*, the independent variable is positive and the dependent variable is negative.

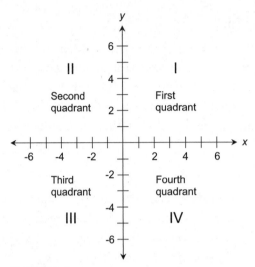

FIGURE 2-3 · We can divide the Cartesian plane into quadrants and label them I (the first quadrant), II (the second quadrant), III (the third quadrant), and IV (the fourth quadrant).

Mathematicians sometimes label the quadrants using Roman numerals, as follows:

- Quadrant I lies at the upper right.
- Quadrant II lies at the upper left.
- Quadrant III lies at the lower left.
- Quadrant IV lies at the lower right.

If a point happens to lie on one of the axes or at the origin, then we don't associate it with any quadrant.

 Still Struggling

You might reasonably ask, "Why do we insist that the increments have identical sizes on both axes in a Cartesian plane?" The answer is simple: Mathematicians define it that way by convention! Nevertheless, not all types of coordinate systems adhere to that strict standard. In a more generalized system called *rectangular coordinates* or the *rectangular coordinate plane*, the two axes can have increments of different sizes (although they must both be linear along their entire length). For example, in a rectangular (but not Cartesian) system, the value on one axis might change by 1 unit for every 5 millimeters, while the value on the other axis changes by 1 unit for every 10 millimeters.

PROBLEM 2-1

Imagine an ordered pair (*x*,*y*), where both variables are nonzero real numbers. Suppose that you've plotted this ordered pair as a point called *P* on the Cartesian plane. Because *x* ≠ 0 and *y* ≠ 0, the point *P* doesn't lie on either axis. What will happen to the location of *P* if you:

- Multiply *x* by −1 and leave *y* the same?
- Multiply *y* by −1 and leave *x* the same?
- Multiply both *x* and *y* by −1?

✔ SOLUTION

If you multiply *x* by −1 but don't change the value of *y*, *P* will move to the opposite side of the *y* axis, but will remain the same distance away from that axis. The point will, in effect, get "reflected" by the *y* axis, moving to the left if *x* is positive to begin with, and to the right if *x* is negative to begin with.

- If *P* starts out in the first quadrant, it will move to the second.
- If *P* starts out in the second quadrant, it will move to the first.
- If *P* starts out in the third quadrant, it will move to the fourth.
- If *P* starts out in the fourth quadrant, it will move to the third.

If you multiply *y* by −1 but leave *x* unchanged, *P* will move to the opposite side of the *x* axis, but will remain the same distance away from that axis. In a sense, *P* will get "reflected" by the *x* axis, moving straight downward if *y* is initially positive and straight upward if *y* is initially negative.

- If *P* starts out in the first quadrant, it will move to the fourth.
- If *P* starts out in the second quadrant, it will move to the third.
- If *P* starts out in the third quadrant, it will move to the second.
- If *P* starts out in the fourth quadrant, it will move to the first.

If you multiply both *x* and *y* by –1, *P* will move diagonally to the opposite quadrant. It will, in effect, get "reflected" by both axes.

- If *P* starts out in the first quadrant, it will move to the third.
- If *P* starts out in the second quadrant, it will move to the fourth.
- If *P* starts out in the third quadrant, it will move to the first.
- If *P* starts out in the fourth quadrant, it will move to the second.

TIP   *If you have trouble envisioning the above-described "point maneuvers," draw a Cartesian plane on a piece of graph paper. Then plot a point or two in each quadrant. Calculate how the variable values change when you multiply either or both of them by –1, and then plot the new points.*

Radial Distance of a Point from the Origin

In the Cartesian plane, the radial (straight-line) distance between any point and the origin depends on both of the numbers in the point's ordered pair.

PROBLEM 2-2

Figure 2-4 shows the point (4,3) plotted in the Cartesian plane. Suppose that we want to find the distance *r* of the point (4,3) from the origin (0,0). How can we do it?

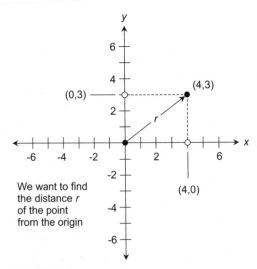

FIGURE 2-4 • Illustration for Problem 2-2.

SOLUTION

We can calculate r using the Pythagorean theorem, which we learned in Chap. 1. Let's restate that theorem using different names for the sides of the triangle. We can call the lengths of the sides x, y, and r as shown in Fig. 2-5. The Pythagorean theorem assures us that

$$x^2 + y^2 = r^2$$

We can use algebra to rewrite this equation as

$$r = (x^2 + y^2)^{1/2}$$

where the 1/2 power of a quantity represents the *nonnegative square root* of that quantity. Now let's make the following point assignments between the scenario of Fig. 2-4 and the scenario of Fig. 2-5:

- The origin in Fig. 2-4 corresponds to the point Q in Fig. 2-5.
- The point (4,0) in Fig. 2-4 corresponds to the point R in Fig. 2-5.
- The point (4,3) in Fig. 2-4 corresponds to the point P in Fig. 2-5.

Continuing with this analogy, we can see the following facts:

- The line segment connecting the origin and (4,0) has length $x = 4$.
- The line segment connecting (4,0) and (4,3) has height $y = 3$.
- The line segment connecting the origin and (4,3) has length r (unknown).

The side of the right triangle having length r constitutes the hypotenuse. Using the Pythagorean formula, we can calculate

$$r = (x^2 + y^2)^{1/2}$$
$$= (4^2 + 3^2)^{1/2}$$
$$= (16 + 9)^{1/2}$$
$$= 25^{1/2}$$
$$= 5$$

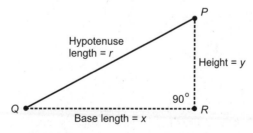

FIGURE 2-5 · Illustration for the solution to Problem 2-2.

We've determined that the point (4,3) lies 5 units away from the origin in Cartesian coordinates, as measured along a straight radial line going out from the origin (0,0) and ending at the point (4,3).

The General Formula

We can generalize the foregoing example to derive a formula for the radial distance of *any* point from the origin in the Cartesian plane. In fact, we can repeat the above explanation almost verbatim, only with a few substitutions.

Consider a point P with coordinates (x_p, y_p). We want to calculate the straight-line radial distance r of the point P from the origin (0,0), as shown in Fig. 2-6. Once again, we use the Pythagorean theorem. Turn back to Fig. 2-5 and follow along by comparing with Fig. 2-6. We can see the following facts:

- The origin in Fig. 2-6 corresponds to the point Q in Fig. 2-5.
- The point $(x_p, 0)$ in Fig. 2-6 corresponds to the point R in Fig. 2-5.
- The point (x_p, y_p) in Fig. 2-6 corresponds to the point P in Fig. 2-5.

These facts are also visually evident:

- The line segment connecting the origin and $(x_p, 0)$ has length $x = x_p$.

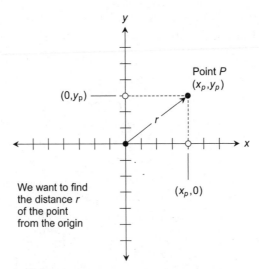

FIGURE 2-6 · Using the Pythagorean theorem, we can derive a formula for the radial distance r of a generalized point $P = (x_p, y_p)$ from the origin.

- The line segment connecting $(x_p, 0)$ and (x_p, y_p) has height $y = y_p$.
- The line segment connecting the origin and (x_p, y_p) has length r (unknown).

The Pythagorean formula tells us that

$$r = (x^2 + y^2)^{1/2}$$
$$= (x_p{}^2 + y_p{}^2)^{1/2}$$

That's all there is to it! The point (x_p, y_p) lies $(x_p{}^2 + y_p{}^2)^{1/2}$ units away from the origin, as we would measure it along a straight line.

Still Struggling

You might ask, "Can the distance of a point from the origin ever turn out as a negative number?" As we've defined it here, the answer is no. If you look at the formula and break down the process in your mind, you'll see why. First, you square x_p, which equals the x coordinate of P. Because x_p is a real number, its square must always equal a *nonnegative* real number. Next, you square y_p, which equals the y coordinate of P. This result must also turn out as a nonnegative real number. Next, you add these two nonnegative real numbers, which must produce another nonnegative real number. Finally, you take the nonnegative square root, getting yet another nonnegative real number! That's the distance of P from the origin. This quantity can't turn out negative in a Cartesian plane whose axes represent real-number variables.

PROBLEM 2-3

Imagine a point $P = (x_p, y_p)$ in the Cartesian plane, where $x_p \neq 0$ and $y_p \neq 0$. Suppose that P lies at a distance of r units away from the origin. What will happen to r if you multiply x_p by -1 and leave y unchanged? If you multiply y_p by -1 and leave x unchanged? If you multiply both x_p and y_p by -1?

 SOLUTION

This problem presents us with a three-part challenge. Let's break each part down into steps and apply the distance formula in each case.

In the first situation, we change the x coordinate of P to its negative. Let's call the new point P_{x-}. It has the coordinates $(-x_p, y_p)$. Let r_{x-} represent the distance of P_{x-} from the origin. Plugging the values into the formula, we obtain

$$r_{x-} = [(-x_p)^2 + y_p^2]^{1/2}$$
$$= [(-1)^2 x_p^2 + y_p^2]^{1/2}$$
$$= (x_p^2 + y_p^2)^{1/2}$$
$$= r$$

In the second situation, we change the y coordinate of P to its negative. This time, let's call the new point P_{y-}. Its has the coordinates $(x_p, -y_p)$. Let r_{y-} represent the distance of P_{y-} from the origin. Plugging the values into the formula, we obtain

$$r_{y-} = [x_p^2 + (-y_p)^2]^{1/2}$$
$$= [x_p^2 + (-1)^2 y_p^2]^{1/2}$$
$$= (x_p^2 + y_p^2)^{1/2}$$
$$= r$$

In the third case, we change both the x and y coordinates of P to their negatives. We can call the new point P_{xy-} with coordinates $(-x_p, -y_p)$. If we let r_{xy-} represent the distance of P_{xy-} from the origin, then

$$r_{xy-} = [(-x_p)^2 + (-y_p)^2]^{1/2}$$
$$= [(-1)^2 x_p^2 + (-1)^2 y_p^2]^{1/2}$$
$$= (x_p^2 + y_p^2)^{1/2}$$
$$= r$$

TIP *We've just demonstrated the fact that we can negate either or both of the coordinate values of a point in the Cartesian plane, and although the point's location will usually change, its distance from the origin will always remain the same.*

Distance between Two Points

We can determine the distance between any two points on a number line by taking the absolute value of the difference between the numbers corresponding to the points. In the Cartesian plane, the process is more complicated.

Setting Up the Problem

Figure 2-7 shows two points, P and Q in the Cartesian plane with coordinates of

$$P = (x_p, y_p)$$

and

$$Q = (x_q, y_q)$$

We want to find the distance d between these points. Let's choose a third point R (which doesn't lie on the line defined by P and Q) and connect P, Q, and R with line segments to get a triangle. The shape of triangle PQR depends on the location of R. If we choose certain coordinates for R, we can get a right triangle with the right angle at R.

With the help of Fig. 2-7, we can envision the coordinates of R. If I travel "straight down" (parallel to the y axis) from P, and if you travel "straight to the right" (parallel to the x axis) from Q, our paths will cross at a right angle when we reach the point with the coordinates (x_p, y_q). Those are the coordinates that R must have, if we want the "base" and the "height" of the triangle to meet at a right angle.

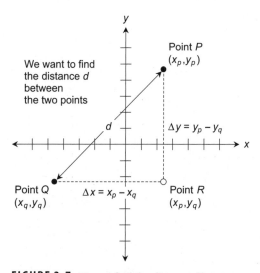

FIGURE 2-7 • We can find the distance d between two points $P = (x_p, y_p)$ and $Q = (x_q, y_q)$ by choosing point R to get a right triangle, and then applying the Pythagorean theorem.

Still Struggling

"Wait!" you say. "Can't we select a point different from *R* as the third vertex of a right triangle along with points *P* and *Q*?" The answer is yes, and Fig. 2-8 shows the situation. If I go "straight up" (parallel to the *y* axis) from *Q*, and if you go "straight to the left" (parallel to the *x* axis) from *P*, we'll meet at a right angle when we reach the coordinates (x_q, y_p). In this case, we can call the right-angle vertex point *S*. We won't use this geometry in the derivation that follows. But we could use it, and the final distance formula would turn out the same.

Dimensions and "Deltas"

Mathematicians use the uppercase Greek letter *delta* (Δ) to stand for "the difference in" or "the difference between." Using this notation, we can state these facts:

- The difference between the *x* values of points *R* and *Q* in Fig. 2-7 equals $x_p - x_q$, or Δx. That's the length of the base of a right triangle.
- The difference between the *y* values of points *P* and *R* in Fig. 2-7 equals $y_p - y_q$, or Δy. That's the height of a right triangle.

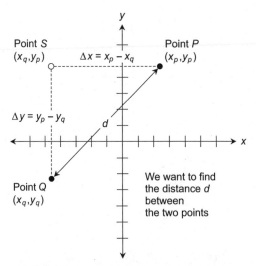

FIGURE 2-8 • Alternative geometry for finding the distance between two points. In this case, the right angle appears at point *S*.

We can see from Fig. 2-7 that the distance d between points Q and P equals the length of the hypotenuse of triangle PQR. Let's find a generalized formula for d using the Pythagorean theorem.

The General Formula

Look back once more at Fig. 2-5 on page 35. The relative positions of points P, Q, and R here are similar to their positions in Fig. 2-7 on page 39. We can define the lengths of the sides of the triangle in Fig. 2-7 as follows:

- The line segment connecting points Q and R has length $\Delta x = x_p - x_q$.
- The line segment connecting points R and P has height $\Delta y = y_p - y_q$.
- The line segment connecting points Q and P has length d (unknown).

The Pythagorean theorem tells us that

$$d = (\Delta x^2 + \Delta y^2)^{1/2}$$
$$= [(x_p - x_q)^2 + (y_p - y_q)^2]^{1/2}$$

 PROBLEM 2-4

Using the formula that we just derived, find the distance d between the following points in the Cartesian plane:

$$P = (-5, -2)$$

and

$$Q = (7, 3)$$

 SOLUTION

When we input the values $x_p = -5$, $y_p = -2$, $x_q = 7$, and $y_q = 3$ into our formula, we get

$$d = [(x_p - x_q)^2 + (y_p - y_q)^2]^{1/2}$$
$$= [(-5 - 7)^2 + (-2 - 3)^2]^{1/2}$$
$$= [(-12)^2 + (-5)^2]^{1/2}$$
$$= (144 + 25)^{1/2}$$
$$= 169^{1/2}$$
$$= 13$$

 **PROBLEM 2-5**

Prove that the distance between two points in the Cartesian plane doesn't depend on the direction of travel.

✔️ **SOLUTION**

When we derived the distance formula a few moments ago, we moved upward and to the right in Fig. 2-7 (from Q to P). When we work with directional displacement, we subtract the starting-point coordinates from the finishing-point coordinates. That's how we got

$$\Delta x = x_p - x_q$$

and

$$\Delta y = y_p - y_q$$

If we travel downward and to the left (from P to Q), we get

$$\Delta^* x = x_q - x_p$$

and

$$\Delta^* y = y_q - y_p$$

when we subtract the starting coordinates from the finishing-point coordinates. These new "star deltas" equal the negatives of the original "plain deltas" because we've done the subtractions in reverse. If we plug the "star deltas" into the derivation for d we worked out a few minutes ago, we can use algebra to get

$$d = (\Delta^* x^2 + \Delta^* y^2)^{1/2}$$
$$= [(-\Delta x)^2 + (-\Delta y)^2]^{1/2}$$
$$= [(-1)^2 \Delta x^2 + (-1)^2 \Delta y^2]^{1/2}$$
$$= (\Delta x^2 + \Delta y^2)^{1/2}$$
$$= [(x_p - x_q)^2 + (y_p - y_q)^2]^{1/2}$$

That's the same distance formula that we got when we traveled upward and to the right from point Q to point P.

TIP   *The foregoing result proves that the direction of travel doesn't matter when we talk about the* distance *between two points in Cartesian coordinates. However, when we talk about* displacement, *as physicists and engineers often do, the direction does matter! We define displacement as traveling a certain distance in a specific direction. Distance values are always nonnegative. Displacement values can be negative or positive.*

Finding the Midpoint

We can find the point midway between two known points on a number line by calculating the *arithmetic mean* (or average value) of the numbers corresponding to the points. In Cartesian xy coordinates, we must make two calculations. First, we average the x values of the two points to get the x value of the point midway between. Then we average the y values of the points to get the y value of the point midway between. Finally we combine the two average values to get an ordered pair.

A "Mini Theorem"

Once again, imagine points P and Q in the Cartesian plane with the coordinates

$$P = (x_p, y_p)$$

and

$$Q = (x_q, y_q)$$

We want to find the coordinates of the *midpoint*—the point that bisects a straight line segment connecting P and Q. As before, we start out by choosing the point R "below and to the right" that forms a right triangle PQR as shown in Fig. 2-9. Imagine a movable point M that we can slide along line segment PQ.

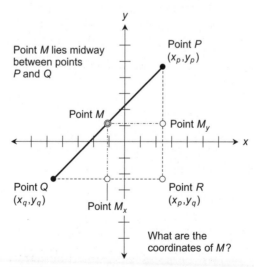

FIGURE 2-9 · We can calculate the coordinates of the midpoint of a line segment whose endpoints we know.

When we draw a perpendicular line from M to side QR, we get a point M_x. When we draw a perpendicular line from M to side RP, we get a point M_y.

Consider the three right triangles MQM_x, PMM_y, and PQR. The laws of basic geometry tell us that these triangles are directly similar, meaning that the lengths of their corresponding sides have identical ratios. According to the definition of direct similarity for triangles, we know these facts:

- Point M_x is midway between Q and R if and only if M is midway between P and Q.

- Point M_y is midway between R and P if and only if M is midway between P and Q.

Now, instead of saying that M stands for "movable point," let's say that M stands for "midpoint." In this case, the x value of M_x (the midpoint of line segment QR) must equal the x value of M, and the y value of M_y (the midpoint of line segment RP) must equal the y value of M.

TIP   *When we prove a "mini theorem" (such as we just did) in order to prove a subsequent, more important result (as we now will), we call the "mini theorem" a lemma.*

The General Formula

We've reduced our Cartesian two-space midpoint problem to two separate number-line midpoint problems. Side QR of triangle PQR runs parallel to the x axis, and side RP of triangle PQR runs parallel to the y axis. We can find the x value of M_x by averaging the x values of Q and R. When we carry out this operation and call the result x_m, we get

$$x_m = (x_p + x_q)/2$$

In the same way, we can calculate the y value of M_y by averaging the y values of R and P. Calling the result y_m, we have

$$y_m = (y_p + y_q)/2$$

We can use the "mini theorem" we finished a few moments ago to conclude that the coordinates of point M, the midpoint of line segment PQ, are

$$(x_m, y_m) = [(x_p + x_q)/2, (y_p + y_q)/2]$$

PROBLEM 2-6

Find the coordinates (x_m, y_m) of the midpoint M between the same two points for which we found the separation distance in the solution to Problem 2-5, that is

$$P = (-5, -2)$$

and

$$Q = (7, 3)$$

SOLUTION

When we plug $x_p = -5$, $y_p = -2$, $x_q = 7$, and $y_q = 3$ into the midpoint formula, we get

$$
\begin{aligned}
(x_m, y_m) &= [(x_p + x_q)/2, (y_p + y_q)/2] \\
&= [(-5 + 7)/2, (-2 + 3)/2] \\
&= (2/2, 1/2) \\
&= (1, 1/2)
\end{aligned}
$$

PROBLEM 2-7

It seems reasonable to suppose that the midpoint between points P and Q shouldn't depend on whether we go from P to Q or from Q to P. We can prove this fact by showing that for all real numbers x_p, y_p, x_q, and y_q, we have

$$[(x_p + x_q)/2, (y_p + y_q)/2] = [(x_q + x_p)/2, (y_q + y_p)/2]$$

SOLUTION

This demonstration is easy, but let's go through it step-by-step to completely follow the logic. For the x coordinates, the rules of basic arithmetic tell us that

$$x_p + x_q = x_q + x_p$$

Dividing each side by 2 gives us

$$(x_p + x_q)/2 = (x_q + x_p)/2$$

Similarly for the y coordinates, we know that

$$y_p + y_q = y_q + y_p$$

Again dividing each side by 2, we get

$$(y_p + y_q)/2 = (y_q + y_p)/2$$

We've shown that the coordinates in the ordered pair on the left-hand side of the original equation precisely equal the corresponding coordinates in the ordered pair on the right-hand side. The ordered pairs are identical, so the midpoint is the same in either direction.

Still Struggling

To find a midpoint of a line segment in Cartesian two-space, you can simply average the coordinates of the endpoints. This method always works if the midpoint lies on a *straight line segment* between the two endpoints. But you might wonder, "How can we find the midpoint between two points along an *arc* connecting those points?" In a situation like that, we must determine the length of the arc. Depending on the nature of the arc, that task can be fairly hard, very difficult, or nigh impossible! Arc-length problems are beyond the scope of this book.

 PROBLEM 2-8

Consider two points in the Cartesian plane, one of which lies at the origin. Show that the coordinate values of the midpoint equal exactly half the corresponding coordinate values of the point that does not lie on the origin.

 SOLUTION

We can plug in (0,0) as the coordinates of either point in the general midpoint formula, and work things out from there. First, let's suppose that point P lies at the origin and the coordinates of point Q are (x_q, y_q). Then $x_p = 0$ and $y_p = 0$. If we call the coordinates of the midpoint (x_m, y_m), we have

$$(x_m, y_m) = [(x_p + x_q)/2, (y_p + y_q)/2]$$
$$= [(0 + x_q)/2, (0 + y_q)/2]$$
$$= (x_q/2, y_q/2)$$

Now, suppose that Q lies at the origin and P has the coordinates (x_p, y_p). In that case, we can derive

$$(x_m, y_m) = [(x_p + x_q)/2, (y_p + y_q)/2]$$
$$= [(x_p + 0)/2, (y_p + 0)/2]$$
$$= (x_p/2, y_p/2)$$

QUIZ

Refer to the text in this chapter if necessary. A good score is eight correct. Answers are in the back of the book.

1. Assuming that we plot the independent variable along the *x* axis and the dependent variable along the *y* axis, Fig. 2-10 shows points in every quadrant of the Cartesian plane *except* the
   A. fourth.
   B. third.
   C. second.
   D. first.

2. What's the distance of the point (−4,5) from the origin in Fig. 2-10? Use a calculator if you need one. Round the answer to three decimal places.
   A. 6.000 units
   B. 6.403 units
   C. 6.667 units
   D. 7.200 units

3. What's the distance of the point (−5,−3) from the origin in Fig. 2-10? Use a calculator if you need one. Round the answer to three decimal places.
   A. 4.000 units
   B. 4.678 units

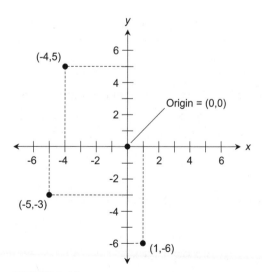

FIGURE 2-10 · Illustration for Quiz Questions 1 through 7.

C. 5.555 units

D. 5.831 units

4. What's the distance of the point (1,−6) from the origin in Fig. 2-10? Use a calcula-
    tor if you need one. Round the answer to three decimal places.

   A. 6.083 units

   B. 6.100 units

   C. 5.900 units

   D. 5.916 units

5. What's the distance between the points (−4,5) and (−5,−3) in Fig. 2-10? Use a
    calculator if you need one. Round the answer to three decimal places.

   A. 7.937 units

   B. 8.000 units

   C. 8.062 units

   D. 8.111 units

6. What's the distance between the points (−5,−3) and (1,−6) in Fig. 2-10? Use a
    calculator if you need one. Round the answer to three decimal places.

   A. 5.196 units

   B. 6.000 units

   C. 6.708 units

   D. 7.222 units

7. What's the distance between the points (1,−6) and (−4,5) in Fig. 2-10? Use a cal-
    culator if you need one. Round the answer to three decimal places.

   A. 12.000 units

   B. 12.083 units

   C. 11.667 units

   D. 11.333 units

8. In Fig. 2-11, the midpoint of line segment L has the coordinates

   A. (−19/4,7/8).

   B. (−4,1).

   C. (−9/2,1).

   D. (−4,7/8).

9. In Fig. 2-11, the midpoint of line segment M has the coordinates

   A. (−2,−9/2).

   B. (−7/3,−37/8).

   C. (−9/4,−17/4).

   D. (−15/8,−9/2).

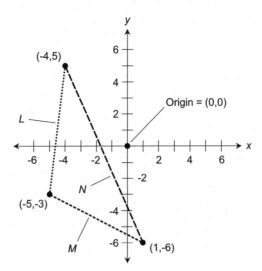

FIGURE 2-11 • Illustration for Quiz Questions 8 through 10.

10. **In Fig. 2-11, the midpoint of line segment *N* has the coordinates**
   A. (−3/2,−1/2).
   B. (−7/4,−1/2).
   C. (−3/2,−5/8).
   D. (−7/4,−11/16).

chapter 3

The Unit-Circle Paradigm

In Chap. 1, we defined the trigonometric functions as ratios between the lengths of the sides of a right triangle. In this chapter, we'll define the same functions in terms of point coordinates on a circle in the Cartesian plane.

CHAPTER OBJECTIVES

In this chapter, you will

- Establish the unit circle in Cartesian coordinates.
- Define the sine and cosine as the coordinates of points on a unit circle.
- Graph the sine and cosine functions.
- Define and graph the tangent function.
- Define and graph the cosecant, secant, and cotangent functions.
- Derive three important relations among pairs of circular functions.

The Unit Circle

In trigonometry, we take a special interest in the circle whose center lies at the origin of the Cartesian plane, and whose radius equals 1 unit. Mathematicians call it the *unit circle*, and represent it with the equation

$$x^2 + y^2 = 1$$

Imagine an angle θ whose vertex lies at the origin. Let's express θ in a counterclockwise rotational sense from the positive x axis, as shown in Fig. 3-1. Suppose that θ defines a ray that intersects the unit circle at a point P, where

$$P = (x_0, y_0)$$

We can portray three *basic circular functions*, also called the *primary circular functions*, of θ in a simple way. But before we do that, let's extend our notion of angles to include negative values, and also angles larger than 2π. Angles constitute the *arguments* (or inputs) of the circular functions.

"Offbeat" versus Standard Angles

In trigonometry, we can reduce any *direction angle*, no matter how extreme, to something nonnegative but less than 2π (360°). Imagine point P starting from (1,0) on the positive x axis, and then revolving around the unit circle so that ray OP rotates with its originating (or back-end) point always at the origin. Even if the ray OP in Fig. 3-1 makes more than one revolution counterclockwise from the x axis, or if it turns clockwise instead of counterclockwise, we can define its direction, at any moment in time, as a counterclockwise angle of at least 0 but less than 2π (360°) relative to the positive x axis.

Think of this situation another way. Point P must always lie somewhere on the unit circle, no matter how many times, or in what direction, ray OP rotates to end up in a particular position. Every point on the circle corresponds to exactly one nonnegative angle less than 2π (360°) going counterclockwise from the positive x axis. Conversely, if we consider the continuous range of angles going counterclockwise over the half-open interval $[0, 2\pi)$ or $[0, 360°)$, we can account for every point on the circle.

We can reduce any "offbeat" direction angle such as $-\pi/4$ or 315° to an equivalent *standard angle* that measures at least 0 but less than 2π (360°) by adding or subtracting some whole-number multiple of 2π (360°). However,

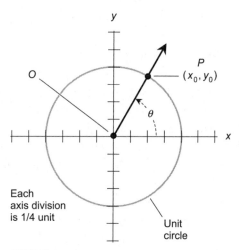

FIGURE 3-1 • The unit circle can serve as the basis for defining trigonometric functions. In this graph, each axis division represents 1/4 unit.

we must be careful with this maneuver. A direction angle specifies orientation only. For example, the orientation of ray OP is the same for an angle of 3π (540°) as for an angle of π (180°), but the larger value carries with it the idea that the ray (also called a *vector*) OP has rotated one and a half times around, while the smaller angle implies that it has undergone only half of a rotation. For our purposes right now, this distinction doesn't matter. But in some situations it does!

 **PROBLEM 3-1**

Suppose that ray OP in Fig. 3-1 rotates 120° clockwise from the positive x axis. What's the equivalent standard angle in degrees?

✔ **SOLUTION**

When we go clockwise, we express angular rotation in the negative sense. We can therefore express the clockwise angle as −120°. To get it into standard form, we must add 360°, getting −120° + 360°, or 240°.

Still Struggling

You'll often encounter negative angles in trigonometry, especially in graphs of functions. Multiple revolutions of objects attain significance in physics and engineering. If you ever read or hear about an angle such as $-\pi/2$ or 5π, you can be sure that it has meaning. The negative value indicates clockwise rotation. An angle of more than 2π (360°) indicates more than one full rotation going around counterclockwise. An angle of less than -2π (−360°) indicates more than one full rotation going around clockwise.

Primary Circular Functions

Look again at Fig. 3-1. Suppose that ray OP points along the positive x axis, and then starts to rotate counterclockwise at a constant speed around its end point O, as if that point is a mechanical hinge. The point P, represented by coordinates (x_0, y_0), revolves around O, following the unit circle.

The Sine Function

Imagine what happens to the value of y_0 (the ordinate of point P) during one complete revolution of ray OP. The ordinate of P starts out at $y_0 = 0$, then increases until it reaches $y_0 = 1$ after P has gone 1/4 of the way around the circle (the ray has turned through an angle of $\pi/2$ or 90°). After that, y_0 begins to decrease, getting back to $y_0 = 0$ when P has gone 1/2 of the way around the circle (the ray has turned through an angle of π or 180°). As P continues in its orbit, y_0 keeps decreasing until the value of y_0 reaches its minimum of −1 when P has gone 3/4 of the way around the circle (the ray has turned through an angle of $3\pi/2$ or 270°). After that, the value of y_0 rises again until, when P has gone completely around the circle, it returns to $y_0 = 0$ when θ reaches 2π (360°). Mathematicians define the value of y_0 as the *sine* of the angle θ. The *sine function* is abbreviated sin, so we can write

$$\sin \theta = y_0$$

PROBLEM 3-2

Based on a visual inspection of Fig. 3-1, what's the sine of $3\pi/2$ (270°)?

SOLUTION

When ray *OP* has rotated through an angle of $3\pi/2$ or 270°, it points straight down along the negative *y* axis. Therefore, ray *OP* intersects the unit circle at the point $(x_0, y_0) = (0, -1)$. The ordinate of this pair equals −1, so

$$\sin(3\pi/2) = \sin 270°$$
$$= -1$$

A Revolving Point

Imagine that you attach a "glow-in-the-dark" ball to the end of a string, and then you swing the ball around at a steady rate of one revolution per second. Suppose that you make the ball circle your head so the path of the ball lies in a horizontal plane. Now imagine that you carry out this experiment in the middle of a flat, open field at night. The glowing ball describes a luminous circle as viewed from high above, as shown in Fig. 3-2A. If a friend stands far away with her eyes exactly in the plane of the ball's orbit, she sees a point of light that oscillates back and forth, from right-to-left and left-to-right, along what looks to her like a straight-line path (Fig. 3-2B). Starting from its rightmost apparent position, the glowing point moves toward the left for 1/2 second, speeding up and then slowing down; then it reverses direction; then it moves toward the right for 1/2 second, speeding up and then slowing down; then turns around again. As seen by your friend, the ball reaches its extreme rightmost position at one-second intervals because its orbital speed equals one revolution per second.

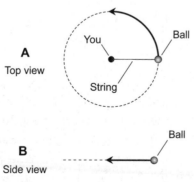

FIGURE 3-2 · Orbiting ball and string. At A, as seen from above; at B, as seen edge-on.

The Sine Wave

If you graph the apparent position of the ball as seen by your friend with respect to time, you'll get a *sine wave* (or *sinusoid*), which forms a coordinate-plane plot of a sine function. Some sine waves "rise higher and lower" (corresponding to a longer string), some appear "flatter" (the equivalent of a shorter string), some look "stretched out" (a slower rate of revolution), and some look "squashed" (a faster rate of revolution). However, these differences don't affect the characteristic of the wave. It's always a sinusoid.

You can whirl the ball around faster or slower than one revolution per second, thereby altering the *frequency* of the sine wave: the number of times a complete wave cycle repeats within a specified interval on the independent-variable axis. You can make the string longer or shorter, thereby adjusting the *amplitude* of the wave: the difference between the extreme values of its dependent variable.

TIP   *Regardless of how you "adjust" the frequency or amplitude of a sine wave, you can always define the result in terms of **uniform circular motion***: *the revolution of a point in a circular orbit at constant speed around a fixed center.*

Graph of a Sine Wave

If we want to graph a sinusoid in the Cartesian plane, the circular-motion analogy can be stated as

$$y = a \sin b\theta$$

where a represents an *amplitude constant* that depends on the radius of the circle, and b represents a *frequency constant* that depends on the revolution rate. As a matter of convention, we always express the angle θ going counterclockwise from the positive x axis. Figure 3-3 portrays a graph of the basic sine function. It's a sinusoid for which $a = 1$ and $b = 1$, and for which we express the angle θ in radians.

The Cosine Function

Look again at Fig. 3-1. Imagine, once again, a ray OP running outward from the origin through point P on the circle. Imagine that at first, the ray points along the positive x axis, and then it rotates steadily in a counterclockwise direction. Now think about what happens to the value of x_0 (the abscissa of point P) during one complete revolution of ray OP. It starts out at $x_0 = 1$, then decreases until it reaches $x_0 = 0$ when θ gets to $\pi/2$ (90°). Then x_0 continues to decrease, getting all the way down to $x_0 = -1$ when θ reaches π (180°). As P continues

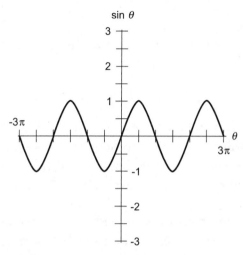

FIGURE 3-3 · Graph of the sine function for values of θ between -3π and 3π. Each division on the horizontal axis represents $\pi/2$ units. Each division on the vertical axis represents $1/2$ unit.

counterclockwise around the circle, x_0 increases. When θ reaches $3\pi/2$ (270°), we get back up to $x_0 = 0$. After that, x_0 increases further until, when P has gone completely around the circle, it returns to $x_0 = 1$ when θ finishes a complete circle and gets to 2π (360°). Mathematicians define the value of x_0 as the *cosine* of the angle θ. The *cosine function* is abbreviated cos, so we can write

$$\cos \theta = x_0$$

PROBLEM 3-3
Based on a visual inspection of Fig. 3-1, what's the cosine of π (180°)?

SOLUTION
When ray OP has rotated through an angle of π (180°), it points straight toward the left along the negative x axis. Therefore, ray OP intersects the unit circle at the point $(x_0, y_0) = (-1, 0)$. The abscissa in this ordered pair equals -1, so

$$\cos \pi = \cos 180°$$
$$= -1$$

The Cosine Wave

We can portray circular motion in the Cartesian plane in terms of the cosine function by means of the equation

$$y = a \cos b\theta$$

where a represents an amplitude constant that depends on the radius of the circle, and b represents a frequency constant that depends on the revolution rate, just as is the case with the sine function. We define or measure θ in the counterclockwise sense from the positive x axis, as always.

TIP   *A cosine wave constitutes a sinusoid, just as a sine wave does. However, for any two constants a and b, the cosine wave is shifted toward the left by 1/4 of a cycle with respect to the sine wave. That shift represents a so-called* **phase angle** *of π/2 or 90°.*

Graph of a Cosine Wave

Figure 3-4 shows a graph of the basic cosine function. It's a *cosine wave* for which $a = 1$ and $b = 1$. The cosine wave in Fig. 3-4 has the same frequency but a 1/4-cycle difference in horizontal position (a phase difference of $\pi/2$ or 90°)

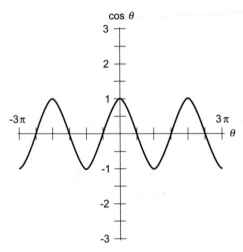

FIGURE 3-4 • Graph of the cosine function for values of θ between -3π and 3π. Each division on the horizontal axis represents $\pi/2$ units. Each division on the vertical axis represents $1/2$ unit.

compared with the sine wave in Fig. 3-3. For those of you who like jargon, electrical engineers call a phase angle of $\pi/2$ or 90° a condition of *phase quadrature*.

The Tangent Function

Once again, refer to Fig. 3-1. We can define the *tangent* (abbreviated tan) of an angle θ in terms of the same ray OP and the same point $P = (x_0, y_0)$ as we used when we defined the sine and cosine functions. We divide the ordinate by the abscissa to get

$$\tan \theta = y_0/x_0$$

We've seen that $\sin \theta = y_0$ and $\cos \theta = x_0$, so we can express the tangent function alternatively as the ratio

$$\tan \theta = \sin \theta/\cos \theta$$

Graph of the Tangent Function

When we plot the tangent function in Cartesian coordinates, we get an interesting graph. It repeats at regular intervals, but it doesn't look like a wave. Unlike the graphs of the sine and cosine functions, a graph of the tangent function "blows up" at certain values of θ as shown in Fig. 3-5. Whenever $x_0 = 0$, the denominator in either of the above formulas attains a value of 0. We can't

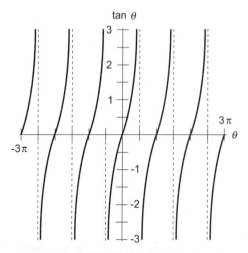

FIGURE 3-5 · Graph of the tangent function for values of θ between -3π and 3π. Each division on the horizontal axis represents $\pi/2$ units. Each division on the vertical axis represents $1/2$ unit.

define the value of the tangent function for any angle θ such that $\cos \theta = 0$. We get a "blowup" whenever θ equals an odd-integer multiple of $\pi/2$.

PROBLEM 3-4

Based on a visual inspection of Fig. 3-1 on page 53, what's the tangent of π (180°)?

✔ SOLUTION

When ray OP has rotated through an angle of π or 180°, it points straight toward the left along the negative x axis. Therefore, ray OP intersects the unit circle at the point $(x_0, y_0) = (-1,0)$, and we have

$$\tan \pi = \tan 180°$$
$$= y_0/x_0$$
$$= 0/(-1)$$
$$= 0$$

PROBLEM 3-5

Based on a visual inspection of Fig. 3-1, what's the tangent of $3\pi/2$ (270°)?

✔ SOLUTION

When ray OP has rotated through an angle of $3\pi/2$ or 270°, it points straight down along the negative y axis, so it intersects the unit circle at $(x_0, y_0) = (0,-1)$. Therefore

$$\tan (3\pi/2) = \tan 270°$$
$$= y_0/x_0$$
$$= -1/0$$

Mathematically, that quotient is undefined.

Singularities

When a function "blows up" as the tangent function does at all the odd-integer multiples of $\pi/2$, we call the function *singular* for the affected values of the input variable, and we call such a "blow-up point" a *singularity*. If you've watched movies about interstellar space travel, maybe you've heard the term *space-time singularity*, referring (supposedly) to a place where the laws of physics break down. In a mathematical singularity, the output value of a function becomes meaningless. In Fig. 3-5, we denote the locations of the singularities by inserting vertical dashed lines that show the curve's *asymptotes*.

Inflection Points

Midway between the singularities, the graph of the tangent function crosses the θ axis, and the sense of the curvature changes. Below the θ axis, the curves are always concave to the right and convex to the left. Above the θ axis, the curves are always concave to the left and convex to the right. Whenever we have a point on a curve where the sense of the curvature reverses, we call that point an *inflection point* or a *point of inflection*. (Some texts spell the word "inflexion.")

Lots of graphs have inflection points. If you're astute, you'll look back in this chapter and notice that the sine and cosine waves also have them. From your algebra courses, you might remember that the graphs of some higher-degree polynomial functions have inflection points.

Still Struggling

Some students wonder if there's any way to define a function at a singularity. If you scrutinize Fig. 3-5 closely, you might suppose that

$$\tan(\pi/2) = \pm\infty$$

where the symbol $\pm\infty$ means *positive or negative infinity*. The graph suggests that the output of the tangent function might attain values of infinity at the singular input points, but we have no formal definition for *infinity* (∞) as a number. Some mathematicians have grappled with the notion of infinity and come up with a way of doing arithmetic with it. Most notable among these people was *Georg Cantor*, a German mathematician who lived from 1845 to 1918. He discovered "multiple infinities" that he called *transfinite numbers*. If you're interested in studying transfinite numbers, try searching the Internet using that term as a phrase.

 **PROBLEM 3-6**

Figure out the value of tan $(\pi/4)$. Don't do any calculations. You should be able to infer this on the basis of geometry alone.

✓ SOLUTION

Draw a diagram of a unit circle, such as the one in Fig. 3-1 on page 53, and place ray *OP* so that it subtends an angle of $\pi/4$ with respect to the positive *x* axis. (That's exactly "northeast" if the positive *x* axis goes "east" and the positive *y* axis goes "north.") Note that the ray *OP* also subtends an angle of $\pi/4$ with respect to the positive *y* axis because the *x* and *y* axes run mutually perpendicular (oriented at an angle of exactly $\pi/2$ with respect to each other), and $\pi/4$ equals half of $\pi/2$. Every point on the ray *OP* lies equally distant from the positive *x* and *y* axes, including the point (x_0, y_0) where the ray intersects the circle. It follows that $x_0 = y_0$. Neither of the coordinate values equal 0, so you know that $y_0/x_0 = 1$. According to the definition of the tangent function, you can conclude that

$$\tan(\pi/4) = y_0/x_0$$
$$= 1$$

Secondary Circular Functions

The three primary circular functions, as defined above, form the cornerstone of trigonometry. However, three more circular functions exist. Their values represent the reciprocals of the values of the primary circular functions.

The Cosecant Function

Imagine the ray *OP* in Fig. 3-1, oriented at a certain angle θ with respect to the positive *x* axis, pointing outward from the origin, and intersecting the unit circle at $P = (x_0, y_0)$. We define the reciprocal of the ordinate, $1/y_0$, as the *cosecant* of the angle θ. The cosecant function is abbreviated csc, so we can write

$$\csc\theta = 1/y_0$$

Because y_0 equals the value of the sine function, the cosecant equals the reciprocal of the sine. For any angle θ, the following equation holds true as long as $\sin\theta \neq 0$:

$$\csc\theta = 1/\sin\theta$$

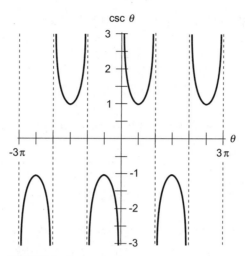

FIGURE 3-6 · Graph of the cosecant function for values of θ between -3π and 3π. Each division on the horizontal axis represents $\pi/2$ units. Each division on the vertical axis represents $1/2$ unit.

We can't define the cosecant of any angle that's any integer multiple of π. The sine of such an angle equals 0, making the cosecant equal to 1/0. Figure 3-6 is a graph of the cosecant function for values of θ between -3π and 3π. The vertical dashed lines denote singularities. A singularity also occurs along the y axis.

PROBLEM 3-7

The cosecant of an angle can never lie within a certain range of values. What range?

SOLUTION

The cosecant equals the reciprocal of the sine. The sine has a span of output values covering the closed interval [−1,1]. In other words, no matter what real-number value the input θ might have, we know that

$$-1 \le \sin \theta \le 1$$

We can split this inequality into two others, dividing the interval in half to get

$$-1 \le \sin \theta \le 0$$

and

$$0 \le \sin \theta \le 1$$

According to the rules of "inequality algebra," it follows that

$$1/\sin \theta \leq -1$$

and

$$1 \leq 1/\sin \theta$$

Because $1/\sin \theta = \csc \theta$, we can rewrite the foregoing two inequalities as

$$\csc \theta \leq -1$$

and

$$1 \leq \csc \theta$$

The cosecant function can never attain any value within the open interval $(-1,1)$.

The Secant Function

Consider the reciprocal of the abscissa, that is, $1/x_0$, in the scenario of Fig. 3-1 on page 53. We define this value as the *secant* of the angle θ. The secant function is abbreviated sec, so we can write

$$\sec \theta = 1/x_0$$

The secant of an angle equals the reciprocal of the cosine. When $\cos \theta \neq 0$, we can always say that

$$\sec \theta = 1/\cos \theta$$

The secant is undefined for any positive or negative odd-integer multiple of $\pi/2$. Figure 3-7 is a graph of the secant function for values of θ between -3π and 3π. Note the input values for which the function is singular (vertical dashed lines).

 PROBLEM 3-8

The secant of an angle can never lie within a certain range of values. What range?

SOLUTION

The secant equals the reciprocal of the cosine. The cosine has a range of output values covering the closed interval $[-1,1]$. That is,

$$-1 \leq \cos \theta \leq 1$$

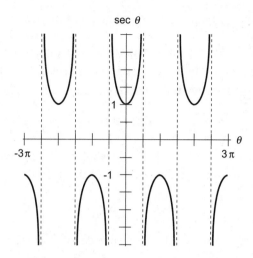

FIGURE 3-7 · Graph of the secant function for values of θ between -3π and 3π. Each division on the horizontal axis represents $\pi/2$ units. Each division on the vertical axis represents $1/2$ unit.

for all real-number input values θ. We can break this fact down into the two statements

$$-1 \leq \cos \theta \leq 0$$

and

$$0 \leq \cos \theta \leq 1$$

According to the rules of "inequality algebra," we can derive

$$1/\cos \theta \leq -1$$

and

$$1 \leq 1/\cos \theta$$

We can rewrite the above pair of inequalities as

$$\sec \theta \leq -1$$

and

$$1 \leq \sec \theta$$

These two statements tell us that the secant function never attains any values in the open interval $(-1,1)$.

The Cotangent Function

Now let's think about the value of x_0/y_0 at the point P where the ray OP crosses the unit circle. We define this ratio as the *cotangent* of the angle θ. The cotangent function is abbreviated cot, so we can write

$$\cot \theta = x_0/y_0$$

Because we already know that $\cos \theta = x_0$ and $\sin \theta = y_0$, we can express the cotangent function in terms of the cosine and the sine as

$$\cot \theta = \cos \theta/\sin \theta$$

The cotangent function also equals the reciprocal of the tangent function; that is,

$$\cot \theta = 1/\tan \theta$$

Whenever $y_0 = 0$, the denominators of all three of the foregoing quotients attain values of 0, so we can't define the cotangent function at any such point. Singularities occur at all integer multiples of π. Figure 3-8 is a graph of the cotangent function for values of θ between -3π and 3π. As in the other examples here, singularities appear as vertical dashed lines.

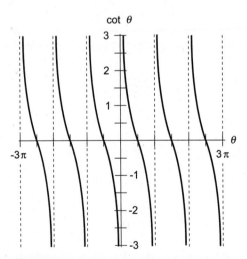

FIGURE 3-8 · Graph of the cotangent function for values of θ between -3π and 3π. Each division on the horizontal axis represents $\pi/2$ units. Each division on the vertical axis represents $1/2$ unit.

TIP   *Now that you know how to define the six circular functions, you might wonder how you can determine the output values of those functions for specific numerical inputs. The easiest way to accomplish this task involves the use of a good scientific calculator. This approach will usually give you an approximation, not an exact value, because the output values of trigonometric functions almost always turn out as irrational numbers. Remember to set the calculator to operate with the input unit you want (either radians or degrees) before you enter anything! The values of the sine and cosine functions never get smaller than −1 or larger than 1. The values of the four other functions can vary wildly. Put a few numbers into your calculator and see what happens when you apply the circular functions to them. When you input a value for which a function has a singularity, you'll get an error message on the calculator.*

 **PROBLEM 3-9**

Figure out the value of cot $(5\pi/4)$. You can solve this problem entirely with geometry, so keep your calculator switched off!

✓ **SOLUTION**

Draw a unit circle on a Cartesian coordinate grid. Orient the ray *OP* so that it subtends an angle of $5\pi/4$ with respect to the positive *x* axis. (That's "southwest" if the positive *x* axis goes "east" and the positive *y* axis goes "north.") Every point on *OP* lies equally distant from the *x* and *y* axes, including (x_0, y_0) where the ray intersects the circle. You can see that $x_0 = y_0$ and both are negative, so the ratio x_0/y_0 must equal 1. According to the definition of the cotangent function, you can conclude that

$$\cot (5\pi/4) = 1$$

PROBLEM 3-10

The Pythagorean formula for the sine and cosine, which we proved in Chap. 1, tells us that

$$\sin^2\theta + \cos^2\theta = 1$$

for all possible values of θ. From this equation, derive the fact that

$$\sec^2\theta - \tan^2\theta = 1$$

for all values of θ where the secant and the tangent are defined.

SOLUTION

We start with the Pythagorean theorem for the sine and cosine. Again, it's

$$\sin^2\theta + \cos^2\theta = 1$$

When we subtract $\sin^2\theta$ from either side, we get

$$\cos^2\theta = 1 - \sin^2\theta$$

We can divide this entire equation by the square of the cosine, as long as we don't allow θ to equal an odd-integer multiple of $\pi/2$. (If θ does attain such a value, then $\cos\theta = 0$, which means that $\cos^2\theta = 0$ and we end up dividing by 0.) Performing the division, we get

$$\cos^2\theta/\cos^2\theta = 1/\cos^2\theta - \sin^2\theta/\cos^2\theta$$

The left-hand side of this equation equals 1 for any value of θ, as long as it's not one of the forbidden values. The first term on the right-hand side equals the reciprocal of the cosine squared, which is the same as the secant squared. The second term on the right-hand side is the ratio of the sine squared to the cosine squared, which is same as the tangent squared. We can therefore simplify the above equation to

$$1 = \sec^2\theta - \tan^2\theta$$

which we can transpose to get the desired result

$$\sec^2\theta - \tan^2\theta = 1$$

PROBLEM 3-11

Once again, consider the Pythagorean formula for the sine and cosine:

$$\sin^2\theta + \cos^2\theta = 1$$

From this equation, derive the fact that

$$\csc^2\theta - \cot^2\theta = 1$$

for all values of θ where the cosecant and the cotangent are defined.

 SOLUTION

Again, we start with the Pythagorean theorem for the sine and cosine, which we state as

$$\sin^2\theta + \cos^2\theta = 1$$

Let's subtract $\cos^2\theta$ from each side. That gives us

$$\sin^2\theta = 1 - \cos^2\theta$$

We can divide this entire equation through by the square of the sine, provided that we don't allow θ to equal an integer multiple of π. (If θ does attain one of those values, then we end up dividing by 0.) When we do that, we get

$$\sin^2\theta/\sin^2\theta = 1/\sin^2\theta - \cos^2\theta/\sin^2\theta$$

The left-hand side of the above equation equals 1 as long as θ isn't one of the forbidden values. The first term on the right-hand side equals the reciprocal of the sine squared; that's the same as the cosecant squared. The second term on the right-hand side equals the ratio of the cosine squared to the sine squared. That's the same as the cotangent squared. We can therefore simplify the above equation to

$$1 = \csc^2\theta - \cot^2\theta$$

which we can transpose to get the sought-after result

$$\csc^2\theta - \cot^2\theta = 1$$

PROBLEM 3-12

Use a drawing of the unit circle to show that $\sin^2\theta + \cos^2\theta = 1$ for angles θ greater than 0 and less than $\pi/2$. (Here's a hint: A right triangle is involved.)

SOLUTION

Figure 3-9 shows the unit circle with θ defined counterclockwise between the positive x axis and a ray emanating from the origin. When θ exceeds 0 but remains less than $\pi/2$, we can construct a right triangle such that:

- The base runs along the x axis extending from the origin to the abscissa of the point where the ray intersects the circle

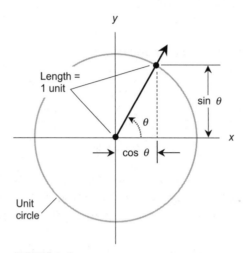

FIGURE 3-9 · Illustration for Problem 3-12.

- The height runs upward at a right angle from the right-hand end of the base to the point where the ray intersects the circle
- The hypotenuse runs outward from the origin to the point where the ray intersects the circle

We can see that the length of the hypotenuse equals the radius of the unit circle. This radius, by definition, equals 1 unit. According to the Pythagorean theorem for right triangles, the square of the length of the hypotenuse equals the sum of the squares of the base length and the height. The lengths of these other two sides are sin θ and cos θ, so it follows that

$$\sin^2\theta + \cos^2\theta = 1$$

PROBLEM 3-13

Use another drawing of the unit circle to show that $\sin^2\theta + \cos^2\theta = 1$ for angles θ greater than $3\pi/2$ and less than 2π. (Here's a hint: This range of angles is equivalent to the range of angles greater than $-\pi/2$ and less than 0.)

 SOLUTION

Refer to Fig. 3-10. It's essentially a mirror image of Fig. 3-9, with the angle θ defined clockwise instead of counterclockwise. When θ is greater than $3\pi/2$ and less than 2π, we can construct a right triangle such that:

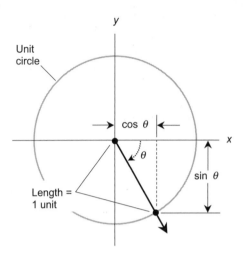

FIGURE 3-10 • Illustration for Problem 3-13.

- The base runs along the *x* axis extending from the origin to the abscissa of the point where the ray intersects the circle
- The height runs downward at a right angle from the right-hand end of the base to the point where the ray intersects the circle
- The hypotenuse runs outward from the origin to the point where the ray intersects the circle

Again, we get a right triangle with a hypotenuse 1 unit long, while the other two sides have lengths of sin θ and cos θ. This triangle, like all right triangles, obeys the Pythagorean theorem. As in the solution to Problem 3-12, we end up with

$$\sin^2 \theta + \cos^2 \theta = 1$$

TIP *Trigonometry abounds with facts that express relationships among the circular functions, some of them quite sophisticated. In the back of this book, you'll find an appendix that outlines a few of these so-called* circular *identities.*

QUIZ

Refer to the text in this chapter if necessary. A good score is eight correct. Answers are in the back of the book.

1. In standard degree measure, an angle of $-\pi/3$ equals
   A. 330°.
   B. 300°.
   C. 270°.
   D. 240°.

2. In standard radian measure, an angle of 855° equals
   A. $\pi/4$.
   B. $\pi/2$.
   C. $3\pi/4$.
   D. π.

3. The graph of the equation $y = 3 \cos 2x$ in Cartesian coordinates, where x can range over the set of values from -6π to 6π, has all of the following characteristics *except one*. Which one?
   A. It contains singularities.
   B. It crosses the x axis several times.
   C. It's a sinusoid.
   D. It's a continuous (unbroken) curve.

4. Suppose that we draw a graph of the circle $x^2 + y^2 = 1$ on the Cartesian plane. Then we choose a point on the circle and draw a ray outward from the origin passing through the point, such that the ray subtends a certain angle relative to the positive x axis. The reciprocal of the point's ordinate equals the
   A. sine of the angle.
   B. cosine of the angle.
   C. secant of the angle.
   D. cosecant of the angle.

5. In the situation described in Question 4, the ratio of the abscissa to the ordinate of the point equals the
   A. secant of the angle.
   B. cosine of the angle.
   C. tangent of the angle.
   D. cotangent of the angle.

6. The graph of the equation $y = \csc x$ in Cartesian coordinates, where x can range over the set of values from -6π to 6π, has *only one* of the following characteristics. Which one?

    A. It contains singularities.
    B. It crosses the *x* axis several times.
    C. It's a sinusoid.
    D. It's a continuous (unbroken) curve.

7. **According to the unit-circle concept, the output value *y* of the function *y* = tan *x* in the Cartesian *x y*-plane can range over the**
    A. closed interval [−1,1].
    B. open interval (−1,1).
    C. entire set of positive real numbers.
    D. entire set of real numbers.

8. **According to the unit-circle concept, the output value *y* of the function *y* = sin *x* in the Cartesian *xy*-plane can range over the**
    A. closed interval [−1,1].
    B. open interval (−1,1).
    C. entire set of positive real numbers.
    D. entire set of real numbers.

9. **According to the unit-circle concept, the output value *y* of the function *y* = cos *x* in the Cartesian *xy*-plane can range over the**
    A. closed interval [−1,1].
    B. open interval (−1,1).
    C. entire set of positive real numbers.
    D. entire set of real numbers.

10. **According to the unit-circle concept, we can't define the cosecant function for an input value of**
    A. 90°.
    B. 180°.
    C. −45°.
    D. 470°.

chapter 4

Mappings, Relations, Functions, and Inverses

We can perform the circular functions "backward" to get the *inverse circular functions*. But first, let's see how mathematical functions derive from *mappings* and *relations*, so that we can identify what constitutes a true function (and what doesn't). We'll need that knowledge to produce sound definitions of the inverse circular functions.

CHAPTER OBJECTIVES

In this chapter, you will

- Define and express mappings between sets of objects.
- Contrast the domain, range, maximal domain, and codomain of a mapping.
- Compare injections, surjections, and bijections.
- See how mappings and ordered pairs give rise to relations and functions.
- Define the inverses of the six circular functions.
- Draw graphs of the inverses of the six circular functions.

What's a Mapping?

Imagine two *sets* (collections) of points, defined by the shaded rectangles in Fig. 4-1. Suppose that you're interested in the *subsets* (portions of the main sets) shown by the hatched ovals. You want to match the points in the top oval with those in the bottom oval. In other words, you want to *map* the elements of one set to the elements of the other set.

Point Matching

Think of Fig. 4-1 as portraying two vans that carry people, some of whom are actively using cell phones to send or receive text messages. The people in one van (call it the upper van) are represented by all the points inside the top rectangle, and the people in the other van (call it the lower van) are represented by all the points inside the bottom rectangle. Within each van, the individual people involved in cell-phone "texting" are represented by points inside the ovals.

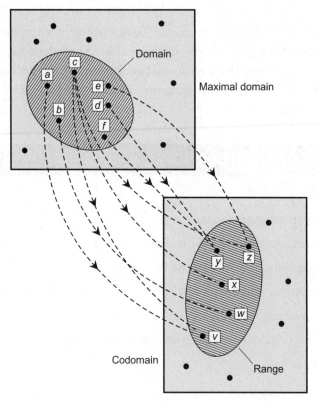

FIGURE 4-1 · A mapping between two sets.

The arrows indicate the direction of the mapping—in this case, the direction in which the messages "travel."

Domain and Range

All of the points actively involved in the mapping shown by Fig. 4-1 lie inside the ovals. We call the contents of the top oval the *domain* of the mapping. Sometimes it's called the *essential domain*. We call the contents of the bottom oval the *range* of the mapping. In this example, the domain has six points and the range has five points. The domain of the mapping is *exactly* the set of people in the upper van who are sending messages. The range is *exactly* the set of people in the lower van who are receiving messages.

Maximal Domain and Codomain

The large rectangles in Fig. 4-1 contain more points than the ovals do. We call the entire contents of the top rectangle the *maximal domain* of the mapping. We call the entire contents of the bottom rectangle the *codomain* of the mapping. Some, but not all, of the points inside the maximal domain and the codomain are actively involved in the mapping. In our example, the maximal domain of the mapping equals the set of all people in the upper van, whether they're sending messages or not. The codomain equals the set of all people in the lower van, whether they're receiving messages or not.

The domain of any mapping forms a subset of the maximal domain. The range of any mapping forms a subset of the codomain. Sometimes, the domain and the maximal domain of a mapping coincide. If that situation were true in the case of Fig. 4-1, then the hatched region would completely fill the top rectangle. Similarly, the codomain and the range of a mapping can sometimes coincide. If that situation were true in the case of Fig. 4-1, then the hatched region would completely fill the bottom rectangle. In any case, the domain of a mapping can never contain any points that lie outside the maximal domain, and the range of a mapping can never contain any points that lie outside the codomain.

Ordered Pairs

When you have established a mapping from the elements of one set to the elements of another set, you can define the mapping in terms of *ordered pairs*. An ordered pair, as you've learned, is an expression in parentheses that contains two items separated by a comma. When we use ordered pairs to describe a mapping, the first item in the pair represents an element of the domain, and the second element represents an element of the range. In the situation shown by Fig. 4-1, the ordered pairs are (a,v), (b,w), (c,v), (c,x), (c,z), (d,y), (e,z), and (f,y).

TIP   *Remember: When you write an ordered pair, don't put any space after the comma, as you would do in an ordinary sequence or a list of set elements.*

Still Struggling

In the mapping that Fig. 4-1 portrays, we don't have a one-to-one correspondence between the points in the domain and the points in the range. Point c in the domain maps to three points in the range. Points v, y, and z in the range each map from two points in the domain. You can imagine that in the upper van, one person transmits text messages to three different people in the lower van. In the lower van, three different people each receive texts from two different senders. "Duplicates" of this sort are "legal" in a general mapping. In some situations, however, we can't allow such "duplicates," as we'll see later in this chapter.

PROBLEM 4-1

Examine Fig. 4-2. Suppose the upper rectangle represents the set of all positive real numbers (or positive reals), and the lower rectangle represents the set of all negative real numbers (or negative reals). Also imagine that the upper oval represents the set of all positive rational numbers (or positive rationals), and the lower oval represents the set of all negative rational numbers (or negative rationals). Now imagine a mapping such that any number x in the upper oval "morphs" into a number y in the lower oval by turning into its additive inverse (getting multiplied by -1). How can we define the ordered pairs in this mapping? What's the domain? The maximal domain? The range? The codomain? What happens in this situation if we want to map a negative real number to something, or if we want to map something to a positive real number?

SOLUTION

We can define the ordered pairs (x,y) as always having the form $(x,-x)$, where x is a positive rational (that is, x is a rational number and $x > 0$). The domain equals the set of all positive rationals. The maximal domain equals

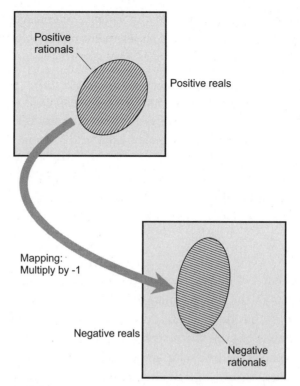

FIGURE 4-2 · A mapping from the positive rational numbers to the negative rational numbers.

the set of all positive reals. The range equals the set of all negative rationals. The codomain equals the set of all negative reals. This mapping doesn't tell us how to map a negative real number to anything. It also fails to tell us how we would map anything to a positive real.

TIP *If you've forgotten the formal definitions of the terms* real number *and* rational number, *take this opportunity to look at one of your old algebra or precalculus textbooks. Review those definitions to make sure that you know the distinction!*

Types of Mappings

Mathematicians have special names for different types of mappings. You should know what these terms mean, even though they might seem strange at first, and even though the distinctions might strike you as the equivalent of "splitting hairs." Imagine two sets of objects, called set X and set Y. Let the variable x

represent an element in set X, and let the variable y represent an element in set Y. We can map the elements of X to the elements of Y in three different ways.

Injection

Figure 4-3 shows a situation in which we map the elements of a set X to the elements of another set Y. This mapping has a domain that forms a subset of X, and a range that forms a subset of Y. Each element x in the domain corresponds to *one and only one* element y in the range. We call a mapping of this type an *injection* or an *injective mapping*. Once in awhile, someone will refer to an injection as a *one-to-one mapping*, or simply *one-to-one*. But that description can mislead because an injection doesn't necessarily involve all the elements of either set X or set Y.

Surjection

The mapping in Fig. 4-4 differs from the one in Fig. 4-3. In the scenario of Fig. 4-4, the elements of the domain map to *all* the elements of Y. The domain forms a subset of X, but the range fills up the entire set Y. We call this type of mapping a *surjection* or a *surjective mapping*. Because a surjection maps elements of the domain completely onto set Y, some mathematicians call it an *onto mapping*, or simply *onto*. A surjection can be one-to-one, but it doesn't have to be. (In the example of Fig. 4-4, it clearly isn't!)

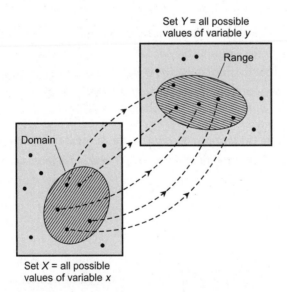

FIGURE 4-3 · An example of an injection. Every element x maps into a single element y, and every element y maps into a single element x.

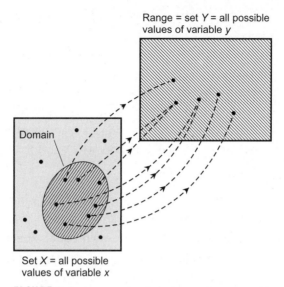

FIGURE 4-4 · An example of a surjection. Every possible element y is accounted for, and maps from at least one element x.

Bijection

Figure 4-5 shows a third type of mapping, called a *bijection*, between two sets X and Y. This type of mapping constitutes an injection that's also a surjection. You might also hear mathematicians (especially set theorists!) say, "A bijection

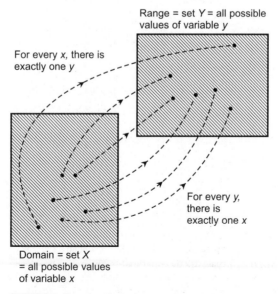

FIGURE 4-5 · An example of a bijection. It acts as both an injection and a surjection.

is both one-to-one and onto." The old-fashioned term for a bijection is *one-to-one correspondence*.

PROBLEM 4-2

Suppose that *X* represents the set of all real numbers *x* that have values larger than 0 but smaller than 1. Suppose that *Y* represents the set of all real numbers *y* that have values strictly larger than 1. Provide an example of an injection from *X* into *Y*. Provide an example of a bijection between *X* and *Y*.

✔ SOLUTION

If we add 1 to any number *x* in set *X*, we get a number *y* in set *Y* that's larger than 1 but smaller than 2. We can write this fact as $y = x + 1$.

This mapping is an injection. The domain equals the whole set *X*, and the range forms a *proper subset* (a subset that doesn't contain all the values in the main set) of *Y* as shown in Fig. 4-6. This mapping is an injection because for any *x* in the domain, there exists *exactly one* (one and only one) *y* in the range, and vice-versa. But the mapping does not go onto the entire set *Y*, so it's not a surjection.

Now let's consider a different mapping. If we take the reciprocal of any number *x* in set *X*, we get a number *y* in set *Y* that's larger than 1. We can write this fact as $y = 1/x$.

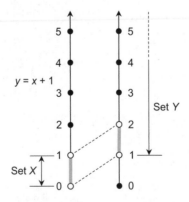

FIGURE 4-6 • An injection from the set *X* of all reals between, but not including, 0 and 1 to the set *Y* of all reals strictly larger than 1. Open circles indicate points not in the domain and range (both of which we show as heavy gray lines).

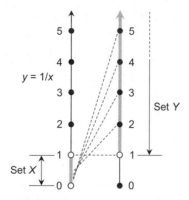

FIGURE 4-7 · A bijection between the set *X* of all reals between 0 and 1 and the set *Y* of all reals strictly larger than 1. Open circles indicate points not included in the domain and range, which appear as heavy gray lines.

No matter what *x* between 0 and 1 we choose, the reciprocal always equals a *unique* (one and only one, or exactly one) real number *y* larger than 1. Therefore, this mapping is an injection. Conversely, no matter what real number *y* larger than 1 we choose, we can always find a unique real number *x* between 0 and 1 that has *y* as its reciprocal. Therefore, our mapping is a surjection. Because it's both an injection and a surjection, we can call the mapping of Fig. 4-7 a bijection.

Examples of Relations

Whenever we deliberately express a mapping in terms of ordered pairs or as an equation in two variables, then that mapping constitutes a mathematical *relation*.

Independent versus Dependent Variable

In a relation, we can represent the elements of the domain and the range as variables. If we say that *x* is a nonspecific element of the domain and *y* is a non-specific element of the range, then *x* acts as the independent variable and *y* acts as the dependent variable. A relation therefore maps values of the independent variable to values of the dependent variable. If we like, we can call *x* the "input variable" and *y* the "output variable," as computer scientists sometimes do.

TIP   *As you learned in Chap. 2, the horizontal axis usually represents the indepen-dent variable in a Cartesian coordinate plane, while the vertical axis usually* *represents the dependent variable. Using this scheme, we can plot almost any relation as a graph on the Cartesian plane. Some relations look like nothing more than "swarms" of points. However, in many cases, the graph of a relation turns out as a line, curve, or other geometric figure. In that case, a graph can reveal things about the relation that we'd never imagine by looking at a set of ordered pairs or an equation.*

An Injective Relation

We can often use equations to represent relations between sets of numbers. We write the dependent variable (such as y) all by itself on the left side of the equality symbol, and then write an expression containing the independent variable (such as x) on the right side. We can get specific ordered pairs by "plugging in" values of the independent variable and then calculating the values of the dependent variable. For example, consider the equation

$$y = x + 2$$

as it applies to all possible real numbers x. When we put specific values of x into this equation and then calculate for y, we obtain results such as the following:

- If $x = -5$ then $y = -3$, so we get $(x,y) = (-5,-3)$
- If $x = -1$ then $y = 1$, so we get $(x,y) = (-1,1)$
- If $x = 0$ then $y = 2$, so we get $(x,y) = (0,2)$
- If $x = 3/2$ then $y = 7/2$, so we get $(x,y) = (3/2,7/2)$
- If $x = 4$ then $y = 6$, so we get $(x,y) = (4,6)$
- If $x = 25$ then $y = 27$, so we get $(x,y) = (25,27)$

This mapping is an injection because for every value of x, we can find exactly one value of y, and vice-versa. We can also call it an *injective relation*.

A Surjective Relation

Suppose that the maximal domain X and the codomain Y of a particular mapping both encompass the entire set of real numbers. Let the essential domain equal the set of all nonnegative real numbers, that is, the set of all x such that $x \geq 0$. Let the range equal the set of all real numbers y (so it coincides with the codomain). Now consider the equation

$$y = \pm(x^{1/2})$$

When we plug specific values of the independent variable x into this equation, we get results such as the following:

- If $x = 1/9$ then $(x,y) = (1/9,1/3)$ or $(1/9,-1/3)$
- If $x = 1/4$ then $(x,y) = (1/4,1/2)$ or $(1/4,-1/2)$
- If $x = 1$ then $(x,y) = (1,1)$ or $(1,-1)$
- If $x = 4$ then $(x,y) = (4,2)$ or $(4,-2)$
- If $x = 9$ then $(x,y) = (9,3)$ or $(9,-3)$
- If $x = 0$ then $(x,y) = (0,0)$

This mapping is clearly not an injection! For every nonzero "input" value of x, we get two "output" values of y. But the mapping goes onto the entire codomain. No matter what real number y we choose, we can square it and get some nonnegative real number x. Therefore, this mapping is a surjection. We can also call it a *surjective relation*.

A Bijective Relation

Let's modify the relation in the preceding section by restricting the codomain and range Y to the set of nonnegative real numbers. Then we can represent our relation as

$$y = x^{1/2}$$

When we see no minus or plus/minus sign in front of an expression raised to the 1/2 power, then by convention, the 1/2 power indicates the nonnegative square root alone, and not the negative one. Now we get only one output value y for every input value x. We've declared all negative output values invalid! Following are some of the ordered pairs in this relation:

- If $x = 1/9$ then $(x,y) = (1/9,1/3)$
- If $x = 1/4$ then $(x,y) = (1/4,1/2)$
- If $x = 1$ then $(x,y) = (1,1)$
- If $x = 4$ then $(x,y) = (4,2)$
- If $x = 9$ then $(x,y) = (9,3)$
- If $x = 0$ then $(x,y) = (0,0)$

We have an injection that goes onto the entire codomain, so it's both injective and surjective. The equation

$$y = x^{1/2}$$

represents a *bijective relation* within the set of nonnegative reals.

TIP   *No matter what nonnegative real number x we plug into the foregoing rela-tion, we get a unique nonnegative real number y out of it. Things also work the opposite way: No matter what nonnegative real number y we want to obtain from this relation, we can find a unique nonnegative real number x that we can "plug in" to get it.*

Examples of Functions

We define a *function* as a relation in which every element in the domain has *at most* one element in the range. In other words, for every value of the indepen-dent variable that we "plug in," we'll never get more than one "output" value for the dependent variable. Nevertheless, a single value of the dependent vari-able might map from two, three, four, or more values of the independent vari-able—even infinitely many—and we can still have a legitimate function. Let's look at three examples of mathematical functions that meet the requirements of this definition.

Add 1 to the Input

Consider a relation in which x represents the independent variable and y rep-resents the dependent variable, and for which the domain and range both encompass the entire set of real numbers. Let's define it by the equation

$$y = x + 1$$

This relation constitutes a true function between x and y because we never get more than one value of y for any value of x. In fact, for every value of x, there exists exactly one value of y. This function is bijective because it maps values of x onto the entire set of real numbers, and it's also one-to-one.

Mathematicians name functions by giving them letters of the alphabet such as f, g, and h. In this notation, we replace the dependent variable by writing down the function letter followed by the independent variable in parentheses. If we want to express the foregoing function in this form, we can write

$$f(x) = x + 1$$

and read the expression aloud as "f of x equals x plus 1."

Square the Input

Let's look at another simple relation. Suppose that v represents the indepen-dent variable and w represents the dependent variable. Further imagine that the domain spans the entire set of reals, but the range is limited to the set of non-negative reals. We define our relation with the equation

$$w = v^2$$

This relation constitutes a function. If we call it g, we can write

$$g(v) = v^2$$

For every value of v in the domain of g, there's exactly one value of w, which we can also call $g(v)$, in the range. But the reverse situation fails! For every nonzero value of w in the range of g, we can find two values of v in the domain. These two values are always negatives of each other. For example, if $w = 49$, then $v = 7$ or $v = -7$. This duplicity does not necessarily pose a problem; a relation can be *many-to-one* and still qualify as a true function. The trouble happens when a relation is *one-to-many*. Then it can't act as a function.

   Our function g is not injective because it's two-to-one except when $v = 0$. Therefore, it can't be bijective. The function g is surjective, however, because we can account for every possible value in its range (the set of nonnegative reals). In formal language we say, "For any nonnegative real number w in the range of g, there exists at least one v in the domain such that $g(v) = w$".

Cube the Input

Here's another relation. This time, let's call the independent variable t and the dependent variable u. The domain and range both cover the entire set of reals. We express our relation as

$$u = t^3$$

This relation qualifies as a function. If we call it h, then we can write it as

$$h(t) = t^3$$

For every value of t in the domain of h, there exists exactly one value of u in the range. The reverse also holds true: For every value of u in the range of h, there exists exactly one t in the domain. This function is injective. It maps onto the entire range, so it's surjective as well. Therefore, h constitutes a bijection.

   TIP *As is the case with relations, drawing a graph can help you see how a function maps the values of its independent variable (the elements of its domain) to the values of its dependent variable (the elements of its range).*

Inverses of Circular Functions

With any relation, you can transpose the values of the independent and dependent variables while leaving their names the same. You can also transpose the domain and the range. When you do both of those things, you get another relation known as the *inverse relation* (or simply the *inverse* if the context is clear). You can denote the inverse of a relation by writing a superscript −1 after its name. If you have $f(x)$, for example, you can write its inverse as $f^{-1}(x)$. The inverse of a function sometimes turns out to be a true function, but not always! Each of the circular functions that we've learned about so far has an inverse: a relation that "undoes" whatever the original function does. If we restrict the domain and the range to certain spans, we can get a true inverse function for any of the circular functions.

How to "Undo" a Function

Let's elaborate on what we mean by the term *inverse function*, or by the expression *the inverse of a function*. In general terms, the inverse of a function, if it exists, exactly "undoes" what the original function does. We'll define this notion more formally in a moment, but first, we must clarify a couple of notational details.

When a function f has an inverse, we can denote it by following its name with a superscript, so we call the inverse f^{-1}. This superscript looks like an exponent, but that's not how it behaves! The function f^{-1} is not the same thing as the −1 power of f, which would equal the reciprocal of f. If you see $f^{-1}(x)$ written somewhere, it means the inverse function of f applied to the variable x. It does not mean $1/[f(x)]$.

Here's the formal definition. Suppose we have a function called f. The inverse function, which we write as f^{-1}, is a function such that

$$f^{-1}[f(x)] = x$$

for all values of x in the domain of f, and

$$f[f^{-1}(y)] = y$$

for all values of y in the range of f. The function f^{-1} "undoes" what f does, and the function f "undoes" what f^{-1} does. If we apply a function to some value of a variable x and then apply the function's inverse to that, we get x back again. If we apply the inverse of a function to some value of a variable y and then apply the original function to that, we get y back again.

Sometimes, we'll find that a function f has an inverse f^{-1} such that we can simply turn f "inside-out" and get its inverse without worrying about whether this scheme will work for all the values in the domain and range of f. But in many cases, things aren't so simple, and we must impose restrictions on a function in order to define an inverse function. Let's look at an example.

Square versus Square Root

Figure 4-8 is a graph of a simple function $f(x) = x^2$. When we plot values of $f(x)$ along the y axis of a Cartesian xy-plane, we get a graph of the equation $y = x^2$. This curve has a shape familiar to anyone who has taken first-year algebra. It's a *parabola* opening upward, with the *vertex* (extreme point) at the origin.

What do you think constitutes the inverse function of f? You might at first say "The square root." If you say that, you're correct, but only to a partial extent! Try graphing the parabola with the x and y variables interchanged. You'll end up plotting the curve for the equation $x = \pm y^2$ in that case, and you'll get Fig. 4-9. This graph a parabola with exactly the same shape as the one for the equation $y = x^2$, but because we've transposed the roles of the variables x and y (and their axes), the parabola has undergone a 90° clockwise turn, so it appears to "lie on its side." This graph portrays a perfectly good mathematical relation, and it also happens to portray a true function that maps values of y to values of x. But it's not a true function that maps values of x to values of y. Let's call this relation $g(x) = \pm x^{1/2}$. If we make the mistake of treating g as a true function,

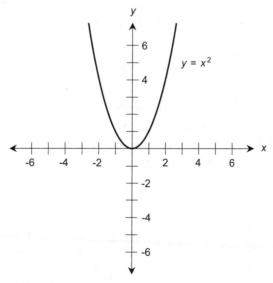

FIGURE 4-8 · The relation $y = x^2$ is a true function of x.

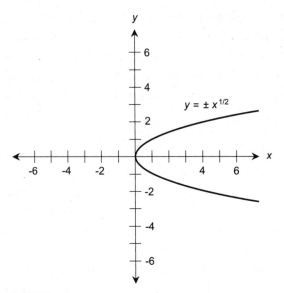

FIGURE 4-9 · The relation $y = \pm x^{1/2}$ does not constitute a true function.

we'll end up with some values of x for which g has no y value (that's okay), and some values of x for which g has two y values (that's *not* okay).

We can modify g so that it behaves as a legitimate function if we're willing to require that y never attain negative values. Alternatively, we can require that the y values never venture into positive territory. Figure 4-10 shows the graph of $y = x^{1/2}$, with the restriction that $y \geq 0$. With this restriction, there exists no abscissa (x value) that has more than one ordinate (y value).

Arc What?

When we want to denote the inverse of a circular function, we can take the standard abbreviation, capitalize it, and add a superscript −1 after it. Alternatively, we can write "Arc" in front of the standard abbreviation. An initial capital letter denotes the fact that we're talking about a true function, and not merely a relation.

- The inverse function for the sine is the Arcsine. If we operate on some variable x, we denote the Arcsine of x as $\mathrm{Sin}^{-1}(x)$ or $\mathrm{Arcsin}(x)$.
- The inverse function for the cosine is the Arccosine. If we operate on some variable x, we denote the Arccosine of x as $\mathrm{Cos}^{-1}(x)$ or $\mathrm{Arccos}(x)$.
- The inverse function for the tangent is the Arctangent. If we operate on some variable x, we denote the Arctangent of x as $\mathrm{Tan}^{-1}(x)$ or $\mathrm{Arctan}(x)$.

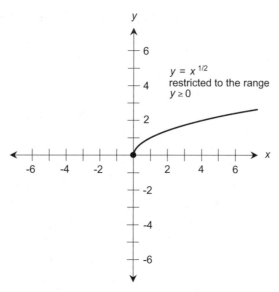

FIGURE 4-10 • The relation $y = x^{1/2}$ behaves as a true function if we restrict y to nonnegative values.

- The inverse function for the cosecant is the Arccosecant. If we operate on some variable x, we denote the Arccosecant of x as $\text{Csc}^{-1}(x)$ or $\text{Arccsc}(x)$.
- The inverse function for the secant is the Arcsecant. If we operate on some variable x, we denote the Arcsecant of x as $\text{Sec}^{-1}(x)$ or $\text{Arcsec}(x)$.
- The inverse function for the cotangent is the Arccotangent. If we operate on some variable x, we denote the Arccotangent of x as $\text{Cot}^{-1}(x)$ or Arccot (x).

TIP   *The sine, cosine, tangent, cosecant, secant, and cotangent functions require special restrictions if we want the inverses to behave as true functions. Figures 4-11 through 4-16, which we'll see in a few moments, take these constraints into account.*

Use (and Misuse) of the −1 Superscript

When using −1 as a superscript in trigonometry, we must exercise caution. Ambiguity, or even nonsense, can result from improper usage. The expression $\text{Sin}^{-1} x$ does not mean the same thing as the expression $(\sin x)^{-1}$. The former refers to the inverse sine or Arcsine function; the latter means the reciprocal of the value that the sine function produces, that is, $1/(\sin x)$. These two operations produce dramatically different results in calculations! If you have any doubt about that fact, "plug in" a few specific values of x to Arcsin x. Then input the same values of x to $(\sin x)^{-1}$ and compare the results.

The foregoing distinction brings up an inconsistency in mathematical syntax. It's customary to write $(\sin x)^2$ as $\sin^2 x$. But don't try that with the exponent -1, for the reason just demonstrated. You might wonder why we should have to treat the numbers 2 and -1 so much differently when we use them as superscripts in trigonometry. I can't come up with a good answer to that question. I can only say that we must follow mathematical protocol in trigonometry (just as we follow traffic laws when we drive our cars). Otherwise we risk misleading our readers (or smashing up our cars!).

TIP *If you ever suspect that the use of a certain notation or expression might produce confusion, don't use it. Use something else, even if it looks clumsy. It's better to look clumsy and state your case correctly, than to appear elegant but say something wrong.*

Still Struggling

You might ask, "Can any function act as its own inverse?" The answer is yes! The function $f(x) = x$ forms its own inverse. In this case, the domain and range both span the entire set of real numbers. If $f(x) = x$, then $f^{-1}(y) = y$. To verify this fact, you can check to verify that the function "undoes its own action," and that this "undoing operation" works both ways. That's an almost trivial task! You get

$$f^{-1}[f(x)] = f^{-1}(x)$$
$$= x$$

and

$$f[f^{-1}(y)] = f(y)$$
$$= y$$

 PROBLEM 4-3

Find another function (besides the identity function described above) that acts as its own inverse.

SOLUTION

Consider $g(x) = 1/x$, with the restriction that the domain and range can attain any real-number value except 0. This function serves as its own inverse; that is,

$$g^{-1}(x) = 1/x$$

To prove this fact, we note that

$$g^{-1}[g(x)] = g^{-1}(1/x)$$
$$= 1/(1/x)$$
$$= x$$

and

$$g[g^{-1}(y)] = g(1/y)$$
$$= 1/(1/y)$$
$$= y$$

for all nonzero real numbers *x* and *y*.

PROBLEM 4-4

Find a function that has no inverse function.

 **SOLUTION**

Consider the function $h(x) = 3$ for all real numbers *x*. If we try to "work" this function in reverse, we must set $y = 3$ in order for $h^{-1}(y)$ to mean anything. The domain of h^{-1} contains the number 3 and nothing else. When we attempt to calculate the value of $h^{-1}(3)$, we get "all the real numbers at once." Clearly, h^{-1} does not constitute a function. It produces infinitely many "output" values from a single "input" value!

Graphs of Inverse Circular Functions

Now that we know what constitutes an inverse function, we can look at the graphs of the circular inverses, placing restrictions on their domains and ranges so that they behave as legitimate functions.

Graph of the Arcsine Function

Figure 4-11 shows a graph of the function $y = \text{Arcsin } x$ (or $y = \text{Sin}^{-1} x$) with its domain restricted to values between and including −1 and 1 (that is, $-1 \le x \le 1$). We restrict the range to values of *y* between and including −90° and 90° ($-\pi/2$ and $\pi/2$). We call numbers within the limited range interval the *principal values*

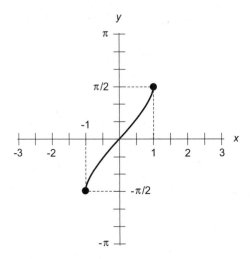

FIGURE 4-11 • Graph of the Arcsine function for
$-1 \le x \le 1$.

of the Arcsine function. We call the entire set of principal values the *principal branch*.

Graph of the Arccosine Function

Figure 4-12 shows a graph of the function $y = \text{Arccos } x$ (or $y = \text{Cos}^{-1} x$) with its domain restricted to values between and including –1 and 1 (that is, $-1 \le x \le 1$).

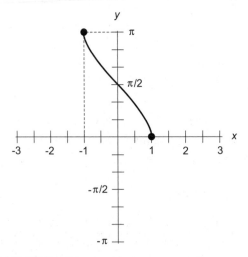

FIGURE 4-12 • Graph of the Arccosine function
for $-1 \le x \le 1$.

We restrict the range to principal values of y between and including $0°$ and $180°$ (0 and π).

Graph of the Arctangent Function

Figure 4-13 shows a graph of the function $y = \text{Arctan } x$ (or $y = \text{Tan}^{-1}x$). The domain encompasses the entire set of real numbers; we don't have to restrict x at all. We confine the range to principal values of y between but not including $-90°$ and $90°$ ($-\pi/2$ and $\pi/2$).

Graph of the Arccosecant Function

Figure 4-14 shows a graph of the function $y = \text{Arccsc } x$ (or $y = \text{Csc}^{-1}x$) with its domain restricted to values less than or equal to -1 or greater than or equal to 1 (that is, $x \leq -1$ or $x \geq 1$). We restrict the range to principal values of y between, and including, $-90°$ and $90°$ ($-\pi/2$ and $\pi/2$), with the exception of 0 (in either degrees or radians).

Graph of the Arcsecant Function

Figure 4-15 shows a graph of the function $y = \text{Arcsec } x$ (or $y = \text{Sec}^{-1}x$) with its domain restricted to values less than or equal to -1 or greater than or equal to 1 (that is, $x \leq -1$ or $x \geq 1$). We restrict the range to principal values of y between, and including, $0°$ and $180°$ (0 and π), with the exception of $90°$ ($\pi/2$).

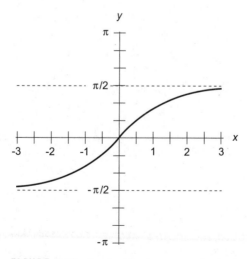

FIGURE 4-13 · Graph of the Arctangent function for $-3 \leq x \leq 3$.

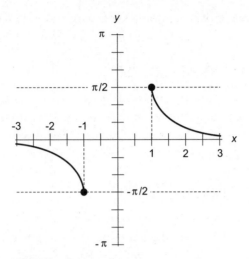

FIGURE 4-14 • Graph the Arccosecant function for $x \le -1$ and $x \ge 1$.

Graph of the Arccotangent Function

Figure 4-16 shows a graph of the function $y = \text{Arccot } x$ (or $y = \text{Cot}^{-1}x$). Its domain encompasses the entire set of real numbers, so we don't have to place any restrictions on x whatsoever. We restrict the range to values of y between, but not including, 0° and 180° (0 and π).

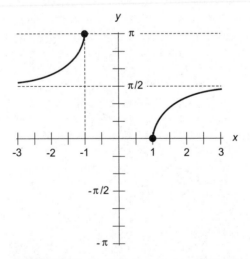

FIGURE 4-15 • Graph of the Arcsecant function for $x \le -1$ and $x \ge 1$.

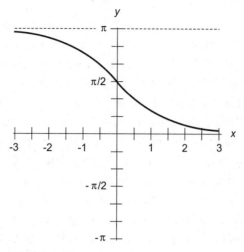

FIGURE 4-16 · Graph of the Arccotangent function for −3 ≤ x ≤ 3.

Still Struggling

Remember: A true function never maps a single value of the independent variable to more than one value of the dependent variable. You can use this fact to determine whether or not a given relation constitutes a function by looking at its graph and conducting a little test. Draw a vertical line somewhere on the graph. ("Vertical" in this context means "parallel to the dependent-variable axis.") Imagine sliding this line to the right and left in the coordinate plane. Sometimes (maybe all the time) the vertical line will intersect the graph of the relation. For the relation to qualify as a true function, the movable vertical line must *never* intersect the graph at more than one point. It's okay if points or "zones" exist where the vertical line fails to intersect the graph. You can call this process the *vertical-line test for a function.*

PROBLEM 4-5

In the preceding six definitions, we've placed well-defined constraints on the domains and ranges of the relations in order to make them act as true functions. Do these restrictions represent the only ones that can work?

SOLUTION

No. We can change the ranges to span other portions of the dependent-variable axis, and we'll still end up with true functions. However, we must use caution if we take this action, making certain that we (1) don't accidentally assign more than one "output" value to any single "input" value, and (2) take all possible values for the domain into account. Generally we can't change the restrictions on the domain, but we can alter the range. As an example, Fig. 4-17 shows three different ways that we can restrict the values of the arcsine relation to obtain a true function. The solid black curve shows the conventional restriction, which gives us the principal values of the Arcsine function. The solid gray curves show two nonstandard alternatives for forcing the arcsine relation to behave as a function. The entire curve portrays the arcsine relation in general. (Note the lowercase "a" when we talk about the arcsine relation instead of the Arcsine function.)

PROBLEM 4-6

Suppose that we restrict the domains or ranges of the inverse circular functions even more severely than Figs. 4-11 through 4-16 suggest. Will the results constitute true functions?

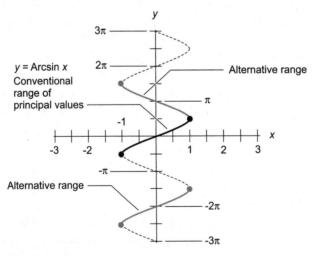

FIGURE 4-17 • Illustration for Problem 4-5.

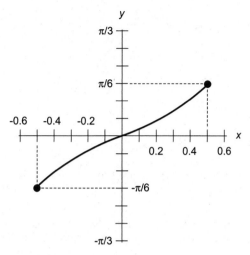

FIGURE 4-18 · Illustration for Problem 4-6.

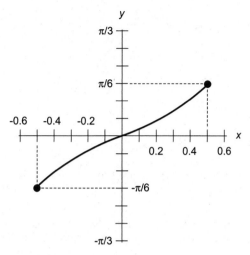SOLUTION

Technically, the answer is yes. However, the resulting functions won't "tell the whole story" because they'll fail to cover all of the possible values. Figure 4-18 shows an example in which we restrict the arcsine relation to a greater extent than we did in the standard graph of Fig. 4-11 on page 94. The vertical-line test will tell us that Fig. 4-18 portrays a function. However, it doesn't portray the entire Arcsine function because it doesn't account for all possible values of that function.

QUIZ

Refer to the text in this chapter if necessary. A good score is eight correct. Answers are in the back of the book.

1. What's the value of Arcsin 90°?
   A. 1
   B. 0
   C. $-\pi/2$
   D. We can't define it; the inverse trigonometric functions operate on plain real numbers, not angles.

2. Examine Fig. 4-17 on page 98 once again. According to this graph, we can modify the arcsine relation to make it behave as a true function if we restrict its range to the interval
   A. $[\pi/2, 3\pi/2]$.
   B. $[-\pi/2, 3\pi/2]$.
   C. $[-\pi, \pi]$.
   D. $[0, \pi]$.

3. Which of the following functions acts as its own inverse function when we restrict the domain to the set of all real numbers except 0?
   A. $f(x) = x$
   B. $g(y) = 1/y$
   C. $h(z) = -z$
   D. All of the above

4. The domain of the sine function extends over the set of
   A. all real numbers.
   B. positive real numbers only.
   C. all real numbers between and including −1 and 1.
   D. all real numbers between and including $-\pi/2$ and $\pi/2$.

5. Which of the following relations, if any, constitutes a true function of x if we place no restrictions on the domain or range?
   A. $x = \cos y$
   B. $x = y^2$
   C. $x = 2y$
   D. None of the above

6. Suppose that we find a mapping between two variables x and y, allowing both variables to extend over the entire set of real numbers, such that every value of x "morphs" to exactly one value of y, and every value of y "morphs" to exactly one value of x. Technically, this mapping constitutes

    A. an injection.
    B. a surjection.
    C. a bijection.
    D. All of the above

7. **The domain of the Arctangent function extends over the set of**
    A. all real numbers.
    B. all real numbers between and including −1 and 1.
    C. all real numbers between and including $-\pi/2$ and $\pi/2$, except 0.
    D. all real numbers except those between and not including −1 and 1.

8. **The range of the Arccosecant function extends over the set of**
    A. all real numbers.
    B. positive real numbers only.
    C. all real numbers except 0.
    D. all real numbers except those between and not including −1 and 1.

9. **Which of the following statements is false?**
    A. The domain of a mapping always forms a subset of the maximal domain.
    B. The range of a mapping always forms a subset of the codomain.
    C. The domain of a mapping always forms a subset of the range.
    D. The ranges of some mappings extend over the entire set of real numbers.

10. **Which of the following functions has no inverse function, if we place no restrictions on the domains or ranges?**
    A. $f(x) = 3$
    B. $g(y) = 5y + 4$
    C. $h(z) = -3z$
    D. $q(w) = w/2$

chapter **5**

Hyperbolic Functions

The *hyperbolic functions* arise from a *unit hyperbola*, the "inside-out" equivalent of a unit circle in Cartesian coordinates. Hyperbolic functions appear in specialized engineering applications. Maybe you'll never see these functions again after you read this chapter; but if you do, at least they won't take you by surprise!

CHAPTER OBJECTIVES

In this chapter, you will

- See how regions around a unit hyperbola portray the hyperbolic sine and cosine functions.
- Express the six hyperbolic functions in exponential terms.
- Define the six inverse hyperbolic functions in logarithmic terms.
- Graph the hyperbolic functions and their inverses.

The "Hyper Six"

As we've learned, the circular functions operate on geometric angles that quantify rotation going counterclockwise around a circle. The hyperbolic functions operate on so-called *hyperbolic angles* that quantify the areas of enclosed regions adjacent to a unit hyperbola. Mathematicians call such regions *hyperbolic sectors*. Six hyperbolic functions exist, known as the *hyperbolic cosine, hyperbolic sine, hyperbolic tangent, hyperbolic cosecant, hyperbolic secant,* and *hyperbolic cotangent*. In formulas and equations, mathematicians abbreviate them as cosh, sinh, tanh, csch, sech, and coth respectively.

Geometric Definitions of Cosh and Sinh

Figure 5-1 shows a graph of the unit hyperbola in a Cartesian uv-plane, where u represents the independent variable (plotted along the horizontal axis) and v

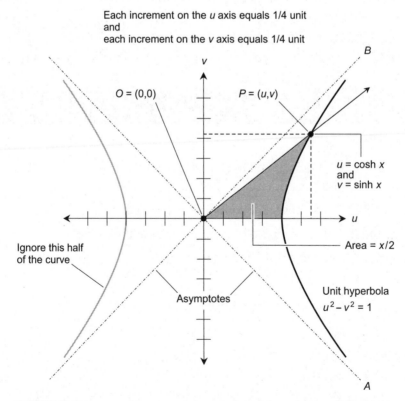

FIGURE 5-1 • The hyperbolic cosine (cosh) and hyperbolic sine (sinh) are based on the areas of enclosed regions adjacent to the right-hand curve in a unit hyperbola.

represents the dependent variable (plotted along the vertical axis). The complete hyperbola has the equation

$$u^2 - v^2 = 1$$

and appears as a pair of curves, one opening up to the right and the other opening up to the left.

Both of the unit hyperbola's curves exhibit *bilateral symmetry* with respect to the horizontal u axis: The bottom halves of the curves form "mirror images" of the top halves. The entire hyperbola also exhibits bilateral symmetry relative to the vertical v axis: The left-hand curve constitutes a "mirror image" of the right-hand curve. In Fig. 5-1, the straight dashed lines through the origin represent *asymptotes*, which form "boundaries" that the curve approaches (but never reaches) as we move away from the origin. In Fig. 5-1, asymptote A is a line with the equation $v = -u$, while asymptote B is a line with the equation $v = u$.

Imagine a ray starting at the origin O = (0,0) and passing through the right-hand curve of the hyperbola at some point $P = (u,v)$. Now consider the area of the region bounded by ray OP, the positive u axis, and the curve. In Fig. 5-1, we see an example; the shaded region represents the area under consideration. If the region lies above the positive u axis, then we call the area positive. If the region lies below the positive u axis, then we call the area negative. Let x represent twice this area (whether it's positive, zero, or negative). We define the hyperbolic cosine of x as

$$\cosh x = u$$

and the hyperbolic sine of x as

$$\sinh x = v$$

for the point $P = (u,v)$ on the right-hand curve of the hyperbola.

Still Struggling

In order to define the hyperbolic cosine and sine functions, we can (and in fact *must*) ignore the left-hand portion of the unit hyperbola. We concern ourselves only with rays OP that emerge from the origin and intersect the right-hand portion of the curve. Remember, the value of a hyperbolic cosine or a hyperbolic sine always derives from the area of an enclosed region, not from the angle that a ray subtends to the positive *u* axis. The domains of the hyperbolic cosine and hyperbolic sine functions both encompass the entire set of real numbers, which we can account for with rays *OP* that emerge between asymptotes *A* and *B* in the right-hand half of the coordinate system (that is, in the first and fourth quadrants).

Exponential Definitions of Cosh and Sinh

We can express the hyperbolic cosine and sine in a completely different way from the geometric scheme we've just seen. These functions also arise from the *base-e exponential functions* that represent powers of a mathematical constant called *e*, which has a variety of interesting properties. For one thing, it's an *irrational number*—a number that we can't express as a ratio of two whole numbers. (In mathematics jargon, the term *irrational* means "impossible to express as a ratio of whole numbers." It does not mean "illogical" or "insane.")

If you have a calculator with a function key marked "e^x" then you can determine the value of *e* to several decimal places. Enter the number 1, and then hit the "e^x" key. If your calculator doesn't have an "e^x" key, it should have a key marked "ln" which stands for *natural logarithm*, and a key marked "inv" which stands for *inverse*. To get *e* from these keys, enter the number 1, and then hit "inv" and "ln" in succession. You should get a number whose first few digits are 2.71828. The full decimal expansion of *e* constitutes a *nonterminating, non-repeating* sequence of digits, as do all irrational numbers.

If you want to determine the value of e^x for some quantity *x* other than 1, you should enter the value *x* and then hit either the "e^x" key or else hit the "inv" and "ln" keys in succession, depending on the type of calculator you have. In order to find e^{-x}, find e^x first, and then find the reciprocal of this by hitting the "$1/x$" key. If *x* represents a real number, we can express the hyperbolic cosine of *x* as

$$\cosh x = (e^x + e^{-x})/2$$

and the hyperbolic sine of *x* as

$$\sinh x = (e^x - e^{-x})/2$$

TIP   *You can have fun trying to pronounce the abbreviations for hyperbolic functions (but not with food in your mouth). Nevertheless, you'll do okay among mathematicians if you name a hyperbolic function in full when you mention it out loud. For example, when you see "sinh," say "hyperbolic sine."*

The Other Four

The remaining four hyperbolic functions arise from the hyperbolic sine and the hyperbolic cosine, in a fashion similar to the way the circular functions relate. We define the hyperbolic tangent (tanh) as

$$\tanh x = \sinh x / \cosh x$$

We define the hyperbolic cosecant (csch) as

$$\operatorname{csch} x = 1/\sinh x$$

We define the hyperbolic secant (sech) as

$$\operatorname{sech} x = 1/\cosh x$$

We define the hyperbolic cotangent as

$$\coth x = \cosh x / \sinh x$$

In terms of exponential functions, we can express the same functions as follows:

$$\tanh x = (e^x - e^{-x})/(e^x + e^{-x})$$
$$\operatorname{csch} x = 2/(e^x - e^{-x})$$
$$\operatorname{sech} x = 2/(e^x + e^{-x})$$
$$\coth x = (e^x + e^{-x})/(e^x - e^{-x})$$

Hyperbolic Function Graphs

Let's look at the graphs of the six hyperbolic functions, based on the definitions we've learned.

Hyperbolic Sine Graph

Figure 5-2 is a graph of the function $y = \sinh x$. Its domain and range both extend over the entire set of real numbers.

Hyperbolic Cosine Graph

Figure 5-3 is a graph of the function $y = \cosh x$. Its domain extends over the whole set of real numbers, and its range encompasses all real numbers y greater than or equal to 1.

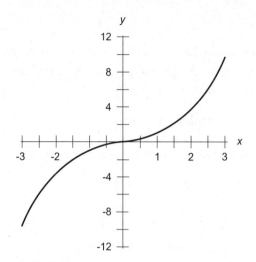

FIGURE 5-2 • Graph of the hyperbolic sine function.

Hyperbolic Tangent Graph

Figure 5-4 is a graph of the function $y = \tanh x$. Its domain spans the entire set of real numbers. The range of the hyperbolic tangent function is limited to the set of real numbers y between, but not including, −1 and 1; that is, $-1 < y < 1$.

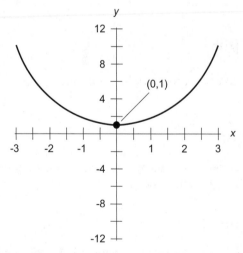

FIGURE 5-3 • Graph of the hyperbolic cosine function.

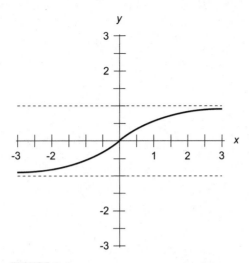

FIGURE 5-4 • Graph of the hyperbolic tangent function.

Hyperbolic Cosecant Graph

Figure 5-5 is a graph of the function $y = \operatorname{csch} x$. Its domain is the set of real numbers x such that $x \neq 0$. The range encompasses all real numbers y such that $y \neq 0$.

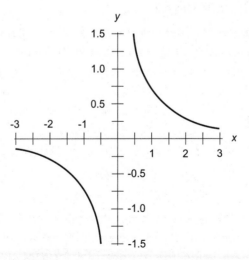

FIGURE 5-5 • Graph of the hyperbolic cosecant function.

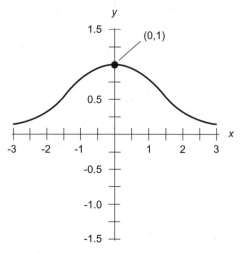

FIGURE 5-6 · Graph of the hyperbolic secant function.

Hyperbolic Secant Graph

Figure 5-6 is a graph of the function $y = \operatorname{sech} x$. Its domain is the entire set of real numbers. Its range is limited to the set of real numbers y greater than 0 but less than or equal to 1; that is, $0 < y \le 1$.

Hyperbolic Cotangent Graph

Figure 5-7 is a graph of the function $y = \coth x$. Its domain is the entire set of real numbers x such that $x \ne 0$. Its range spans the real numbers y less than -1 or greater than 1; that is, $y < -1$ or $y > 1$.

 PROBLEM 5-1

Why does the graph of $y = \operatorname{csch} x$ (shown in Fig. 5-5) "blow up" when $x = 0$? Why can't we define csch 0?

 SOLUTION

Remember that the hyperbolic cosecant (csch) constitutes the reciprocal of the hyperbolic sine (sinh). If $x = 0$, then $\sinh x = 0$, as you can see from Fig. 5-2 (or calculate using the exponential form of the function). As x approaches 0 (written $x \to 0$) from either side, the value of the hyperbolic sine also approaches 0 ($\sinh x \to 0$). Therefore csch x, which equals $1/\sinh x$

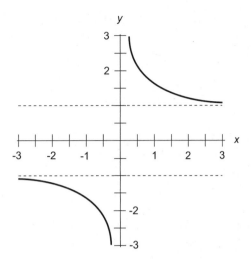

FIGURE 5-7 · Graph of the hyperbolic cotangent function.

and whose graph appears in Fig. 5-5, "blows up positively" as $x \to 0$ from the positive, or right, side (written $x \to 0^+$) and "blows up negatively" as $x \to 0$ from the negative, or left, side ($x \to 0^-$). When $x = 0$, we can't define the reciprocal of the hyperbolic sine because it takes the form of a quotient with 0 in the denominator.

PROBLEM 5-2

What's the hyperbolic cotangent of 0? Express it in two ways.

SOLUTION

This quantity is not defined. We can verify this fact by looking at the graph of the hyperbolic cotangent function (Fig. 5-7). The graph of $y = \coth x$ "blows up" at $x = 0$. It has no y value there. We can also express coth 0 by finding the values of sinh 0 and cosh 0 using the exponential definitions, and then dividing the latter by the former. Remember the formulas

$$\sinh x = (e^x - e^{-x})/2$$

and

$$\cosh x = (e^x + e^{-x})/2$$

If $x = 0$, then $e^x = 1$ and $e^{-x} = 1$. Therefore

$$\sinh 0 = (1 - 1)/2$$
$$= 0/2$$
$$= 0$$

and

$$\cosh 0 = (1 + 1)/2$$
$$= 2/2$$
$$= 1$$

The hyperbolic cotangent equals the hyperbolic cosine divided by the hyperbolic sine, so we have

$$\coth 0 = \cosh 0 / \sinh 0$$
$$= 1/0$$

We can't define this ratio because its denominator equals 0.

Hyperbolic Inverses

Each of the six hyperbolic functions has an inverse relation that we can restrict to obtain a true function. We call them the *hyperbolic Arcsine, hyperbolic Arccosine, hyperbolic Arctangent, hyperbolic Arccosecant, hyperbolic Arcsecant*, and *hyperbolic Arccotangent* functions. In formulas and equations, we abbreviate them as Arcsinh or Sinh^{-1}, Arccosh or Cosh^{-1}, Arctanh or Tanh^{-1}, Arccsch or Csch^{-1}, Arcsech or Sech^{-1}, and Arccoth or Coth^{-1} respectively.

The Natural Logarithm

Mathematicians abbreviate the natural logarithm of x by writing "$\ln x$." This function acts as the inverse of the base-e exponential function. The natural logarithm function and the base-e exponential function "undo" each other.

Suppose that x and v represent real numbers. Let y and u represent positive real numbers. We can write the following two statements relating the natural logarithm function and the natural logarithm base:

- If $e^x = y$, then $x = \ln y$
- If $\ln u = v$, then $u = e^v$

TIP *The natural logarithm function can help you calculate values of the inverse hyperbolic functions, just as the exponential function can help you calculate values of the hyperbolic functions.*

Still Struggling

You can find the natural logarithm of a number using a calculator. Enter the number for which you want to find the natural logarithm, and then hit the "ln" key. Note that only positive quantities have real-number natural logarithms.

Hyperbolic Inverses as Logarithms

You can find hyperbolic inverses of specific quantities using a calculator that has the "ln" function. Following are the expressions for the hyperbolic inverse functions in terms of natural logarithms. The 1/2 power represents the positive square root.

$$\text{Arcsinh } x = \ln[x + (x^2 + 1)^{1/2}]$$

$$\text{Arccosh } x = \ln[x + (x^2 - 1)^{1/2}]$$

$$\text{Arctanh } x = \{\ln[(1 + x)/(1 - x)]\}/2$$

$$\text{Arccsch } x = \ln[x^{-1} + (x^{-2} + 1)^{1/2}]$$

$$\text{Arcsech } x = \ln[x^{-1} + (x^{-2} - 1)^{1/2}]$$

$$\text{Arccoth } x = \{\ln[(x + 1)/(x - 1)]\}/2$$

TIP *Remember to use a capital "A" in the "Arc" prefix to denote the fact that you're writing about true functions, not mere relations.*

Still Struggling

In the third and sixth expressions above, you can consider the denominator 2 as a mathematically exact constant. All of the foregoing formulas are messy, but if you plug in the numbers and take your time doing the calculations, you shouldn't have any trouble with them. Remember how parentheses, square brackets, and curly brackets (also called braces) work in mathematical expressions! (You should have learned the "grouping symbol hierarchy" in your pre-algebra or algebra course.)

Hyperbolic Inverse Graphs

Let's look at the graphs of the six inverse hyperbolic functions, based on the definitions we've learned.

Hyperbolic Arcsine Graph

Figure 5-8 is a graph of the function $y = \text{Arcsinh } x$ (or $y = \text{Sinh}^{-1} x$). Its domain and range both encompass all real numbers.

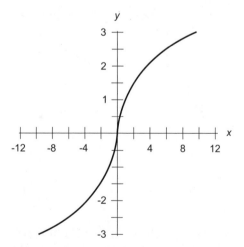

FIGURE 5-8 · Graph of the hyperbolic Arcsine function.

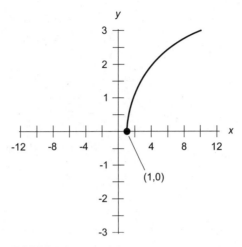

FIGURE 5-9 • Graph of the hyperbolic Arccosine function.

Hyperbolic Arccosine Graph

Figure 5-9 is a graph of the function $y = \text{Arccosh } x$ (or $y = \text{Cosh}^{-1} x$). The domain includes real numbers x such that $x \geq 1$. The range is limited to real numbers y such that $y \geq 0$.

Hyperbolic Arctangent Graph

Figure 5-10 is a graph of the function $y = \text{Arctanh } x$ (or $y = \text{Tanh}^{-1} x$). The domain is limited to real numbers x such that $-1 < x < 1$. The range is the entire set of real numbers.

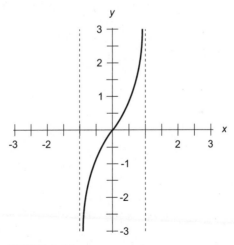

FIGURE 5-10 • Graph of the hyperbolic Arctangent function.

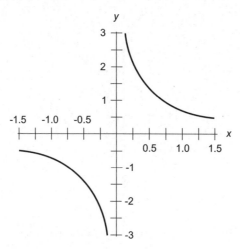

FIGURE 5-11 • Graph of the hyperbolic Arccosecant function.

Hyperbolic Arccosecant Graph

Figure 5-11 is a graph of the function $y = \text{Arccsch } x$ (or $y = \text{Csch}^{-1}x$). Both the domain and the range include all real numbers except 0.

Hyperbolic Arcsecant Graph

Figure 5-12 is a graph of the function $y = \text{Arcsech } x$ (or $y = \text{Sech}^{-1}x$). The domain is limited to real numbers x such that $0 < x \le 1$. The range encompasses all nonnegative real numbers.

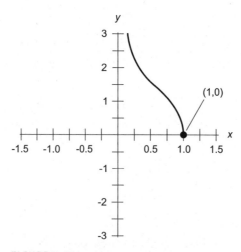

FIGURE 5-12 • Graph of the hyperbolic Arcsecant function.

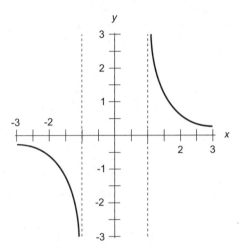

FIGURE 5-13 · Graph of the hyperbolic Arccotangent function.

Hyperbolic Arccotangent Graph

Figure 5-13 is a graph of the function $y = \mathrm{Arccoth}\ x$ (or $y = \mathrm{Coth}^{-1}x$). The domain includes all real numbers x such that $x < -1$ or $x > 1$. The range includes all real numbers except 0.

**PROBLEM 5-3**

What's the value of Arcsinh 0? Use a calculator if you need it.

SOLUTION

When you examine Fig. 5-8, the graph of the hyperbolic Arcsine function, you'll suspect that Arcsinh 0 = 0. You can verify this fact using the formula

$$\mathrm{Arcsinh}\ x = \ln\ [x + (x^2 + 1)^{1/2}]$$

Plugging in 0 for x, you'll get

$$\mathrm{Arcsinh}\ 0 = \ln\ [0 + (0^2 + 1)^{1/2}]$$
$$= \ln\ (0 + 1^{1/2})$$
$$= \ln\ (0 + 1)$$
$$= \ln\ 1$$
$$= 0$$

TIP   *If you've had any experience with logarithms, you don't need a calculator to perform the above calculation because you already know that ln 1 = 0.*

 PROBLEM 5-4

What's the value of Arccsch 1? Use a calculator if you need it. Use the logarithm-based formulas to determine the answer, and express it to three decimal places.

SOLUTION

From the graph in Fig. 5-11, you can guess that Arccsch 1 ought to be a little less than 1. You can get more precise with the help of the logarithmic formula

$$\text{Arccsch } x = \ln [x^{-1} + (x^{-2} + 1)^{1/2}]$$

Plugging in 1 for *x* and calculating, you'll obtain

$$\text{Arccsch } 1 = \ln [1^{-1} + (1^{-2} + 1)^{1/2}]$$
$$= \ln [1 + (1 + 1)^{1/2}]$$
$$= \ln (1 + 2^{1/2})$$
$$= 0.881 \text{ (rounded to three decimal places)}$$

TIP   *In the back of this book, you'll find an appendix that outlines a few basic facts concerning hyperbolic functions, known as* hyperbolic identities.

QUIZ

Refer to the text in this chapter if necessary. A good score is eight correct. Answers are in the back of the book.

1. As *x* starts out at 0 and then increases positively toward 1, the value of Arctanh *x*
   A. approaches 0.
   B. approaches −1.
   C. increases positively without limit.
   D. remains constant.

2. As *x* increases negatively without limit, the value of tanh *x*
   A. approaches −1.
   B. increases negatively without limit.
   C. approaches 1.
   D. increases positively without limit.

3. Using the exponential formula along with a calculator, we can approximate the hyperbolic cosine of 1 (assuming the value 1 to be exact) as
   A. 1.543.
   B. 1.675.
   C. 1.833.
   D. 2.000.

4. The natural logarithm of *e*
   A. equals 0.
   B. equals e^{-1}.
   C. equals 1.
   D. is undefined.

5. The hyperbolic cotangent is equivalent to
   A. the reciprocal of the hyperbolic sine, assuming that the hyperbolic sine doesn't equal 0.
   B. the ratio of the hyperbolic sine to the hyperbolic cosine, assuming that the hyperbolic cosine doesn't equal 0.
   C. the reciprocal of the hyperbolic cosine, assuming that the hyperbolic cosine doesn't equal 0.
   D. the ratio of the hyperbolic cosine to the hyperbolic sine, assuming that the hyperbolic sine doesn't equal 0.

6. We can quantify a hyperbolic angle in terms of
   A. the ratio of two angles in rectangular coordinates.
   B. the displacement along a straight line emanating from the origin.
   C. the area of an enclosed region adjacent to a hyperbola.
   D. the distance between two points on a hyperbola.

7. From the logarithm formulas and with the aid of a calculator, we can deduce that the hyperbolic Arcsine of −4.0000 equals approximately
   A. −10.87.
   B. −2.095.
   C. 3.297.
   D. 2.108.

8. As the value of x becomes larger positively without limit, the value of csch x
   A. also becomes larger positively without limit.
   B. approaches 0.
   C. approaches 1.
   D. becomes larger negatively without limit.

9. How does the hyperbolic sine of 3 compare with the hyperbolic sine of −3?
   A. They're reciprocals of each other.
   B. They add up to 0.
   C. Their ratio equals e.
   D. They're the same.

10. How does the hyperbolic cosine of 3 compare with the hyperbolic cosine of −3?
   A. They're reciprocals of each other.
   B. They add up to 0.
   C. Their ratio equals e.
   D. They're the same.

chapter **6**

Polar Coordinates

The Cartesian plane isn't the only tool for graphing on a flat surface. Instead of moving horizontally and vertically away from the origin to reach a desired point, we can travel in a specified direction straight out from the origin to reach that point. This scheme gives us *polar two-space*, also called the *polar coordinate plane* or simply *polar coordinates*.

CHAPTER OBJECTIVES

In this chapter, you will

- Compare independent-variable and dependent-variable axes in polar coordinates.
- Identify equations that yield lines, circles, and spirals in polar coordinates.
- Convert polar coordinates to Cartesian coordinates.
- Convert Cartesian coordinates to polar coordinates.
- Contrast the navigators' polar system with the mathematicians' polar system.

The Variables

Figure 6-1 shows the basic polar coordinate plane. The independent variable corresponds to an angle θ, called the *direction*, relative to a ray pointing to the right (or "east"), which constitutes the *reference axis*. The dependent variable corresponds to the distance r, called the *radius*, from the origin. We define specific points as ordered pairs of the form (θ, r).

Radius

In the polar coordinate plane, the radial markers form concentric circles. As the circle radius increases, so does the value of r. In Fig. 6-1, we haven't labeled the circles in units, so we can imagine each concentric circle, working outward, as increasing by any number of units we want. For example, each radial division (distance between a pair of adjacent concentric circles) might represent 1 unit, or 5 units, or 10 units, or 100 units. Whatever size increments we choose, we must make sure that they stay the same size all the way out. In other words, we must have a *linear relationship* between the radius coordinate and the actual radius of its circle.

Direction

Mathematicians express polar-coordinate direction angles in radians. We go counterclockwise from a reference axis pointing in the same direction as the

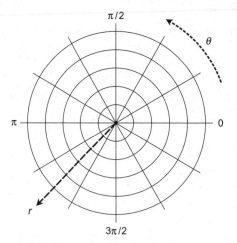

FIGURE 6-1 · The polar coordinate plane. Rays emanating from the origin represent angular increments. Circles centered at the origin represent radial increments.

positive x axis normally goes in the Cartesian xy-plane. The physical angle on the graph must vary in direct proportion to the value of θ. In other words, we must have a linear relationship between the direction coordinate and the actual angle with respect to the reference axis.

TIP *In polar coordinates, we can have nonstandard direction angles. If $\theta \geq 2\pi$, it represents at least one complete counterclockwise rotation from the reference axis. If $\theta < 0$, it represents clockwise rotation from the reference axis. We can also have negative radius coordinates. If we encounter some point for which we're told that $r < 0$, we can multiply r by -1 to make it positive, and then add or subtract π to or from the direction angle. That's something like saying "Walk 100 meters due east" instead of "Walk -100 meters due west."*

Which Variable Is Which?

If you read a lot of mathematics texts and papers, you'll sometimes see ordered pairs for polar coordinates with the radius listed first, and then the angle. Instead of the form (θ, r), you'll see the form (r, θ). In this scheme, the radius represents the independent variable while the direction represents the dependent variable. That works okay, but most people find it easier to imagine the direction as the independent variable with the radius dependent on it.

Imagine an old-fashioned radar display (like the ones shown in war movies made in the 1940s and 1950s). A bright radial ray rotates around a circular screen, revealing targets at various distances. The rotation continues at a steady rate; it's *independent*. Target distances are functions of the direction. Theoretically, a radar display could work with an expanding bright circle instead of a rotating ray, and all of the targets would show up in the same places, but that geometry was never used. In this course, let's use the (θ, r) format for ordered pairs, where θ represents the independent variable and r represents the dependent variable.

Still Struggling

You ask, "How can we write down relations and functions intended for polar coordinates as opposed to those meant for Cartesian coordinates?" It's simple. When we want to denote a relation or function (call it f) in polar coordinates where the independent variable is θ and the dependent variable is r, we write

$$r = f(\theta)$$

We can read this equation in words as "r equals f of θ." When we want to denote a relation or function (call it g) in Cartesian coordinates where the independent variable is x and the dependent variable is y, we can write

$$y = g(x)$$

We can read this statement in words as "y equals g of x."

PROBLEM 6-1

Provide an example of a graphical object that represents a function in polar coordinates when θ represents the independent variable, but not in Cartesian xy coordinates when x represents the independent variable.

✔ SOLUTION

Consider a polar function that maps all input values to the same output value, such as

$$f(\theta) = 3$$

Because $f(\theta)$ gives us an alternative way to define the radius r, this function tells us that $r = 3$. The polar graph forms a circle with a radius of 3 units, centered at the origin. In Cartesian coordinates, the equation of the same circle is

$$x^2 + y^2 = 9$$

where $9 = 3^2$, the square of the radius. If we let y represent the dependent variable and x represent the independent variable, we can rearrange this equation to get

$$y = \pm(9 - x^2)^{1/2}$$

If we say that $y = g(x)$, we can't claim that g constitutes a true function of x. For most values of x (the independent variable) on the circle, we get two values of y (the dependent variable). For example, if $x = 0$, then $y = \pm 3$. We can call g a relation, but we can't call it a function.

Three Basic Graphs

Let's look at the graphs of three generalized equations in polar coordinates. In the Cartesian plane, all equations of these forms produce straight-line graphs. In polar coordinates, only one of them does.

Constant Angle

When we set the direction angle to a numerical constant, we get a simple polar equation of the form

$$\theta = a$$

where a represents a fixed real number. As we allow the value of r to range over the entire set of real numbers, the polar graph of any such equation forms a straight line passing through the origin, subtending an angle of a with respect to the reference axis. Figure 6-2 shows two examples. In these cases, the equations are

$$\theta = \pi/3$$

and

$$\theta = 7\pi/8$$

Constant Radius

When we set the radius to a fixed real-number constant a, we get a polar equation of the form

$$r = a$$

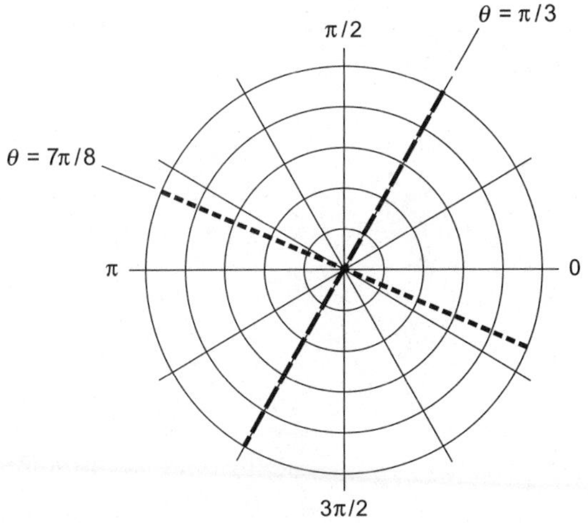

FIGURE 6-2 • When we set the angle constant, we get a straight line that passes through the origin.

As we allow the direction angle θ to rotate through at least one full turn of 2π radians, the graph forms a circle centered at the origin whose radius equals a, as shown in Fig. 6-3. If we allow the angle to span the entire set of real numbers, we trace around the circle infinitely many times, but that redundancy doesn't change the appearance of the graph. Interestingly, the equation $r = -a$ represents the same circle as the equation $r = a$.

Angle Equals Radius Times Positive Constant

Figure 6-4 shows an example of what happens in polar coordinates when we set the radius equal to a positive constant multiple of the angle. We get a pair of "mirror-image spirals."

Imagine a ray pointing from the origin straight out toward the right along the reference axis (labeled 0). In this case $\theta = 0$, so $r = 0$. Suppose that the ray starts to rotate counterclockwise, like the sweep on an old-fashioned military radar screen. The angle *increases positively* at a constant rate. Therefore, the radius also increases at a constant rate, because the radius equals a positive constant multiple of the angle. The resulting graph forms the solid spiral in Fig. 6-4. The *pitch* (or "tightness") of the spiral depends on the value of the constant a in the equation

$$r = a\theta$$

Small positive values of a produce tightly curled-up spirals. Large positive values of a produce loosely pitched spirals.

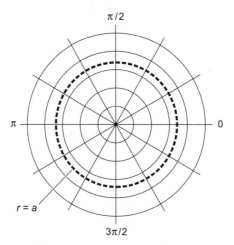

FIGURE 6-3 • When we set the radius constant, we get a circle centered at the origin.

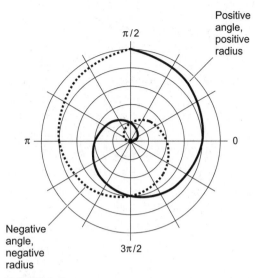

FIGURE 6-4 • When we set the radius to a positive constant multiple of the angle, we get a pair of spirals. Illustration for Problem 6-2.

Now imagine that the ray starts from the reference axis and rotates clockwise instead of counterclockwise. At first, $\theta = 0$, so $r = 0$. As the ray turns, the angle *increases negatively* at a constant rate. The radius also increases negatively at a constant rate, because we multiply the angle by a positive constant. We must therefore plot our points in the opposite direction from the ray. When we do things that way, we get the dashed spiral in Fig. 6-4. The dashed spiral has the same pitch as the solid spiral does, because we haven't changed the value of *a*. The complete graph of the equation contains both spirals together "at the same time."

Angle Equals Radius Times Negative Constant

Figure 6-5 shows an example of what happens in polar coordinates when we set the radius equal to a negative constant multiple of the angle. As in the previous case, we get a pair of spirals, but they're "upside-down" with respect to the situation with a positive constant. To see why things work out that way, you can trace around with rotating, imaginary "radar-sweep" rays, just as you did in Fig. 6-4.

TIP *When you use polar coordinates, you must take care to properly distinguish between positive and negative signs and directions! Positive angles go counter-*

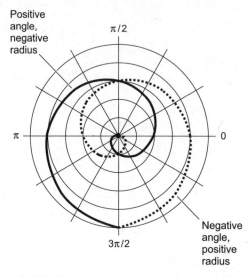

Positive
angle,
negative
radius

π/2

π

0

3π/2

Negative
angle,
positive
radius

FIGURE 6-5 · When we set the radius to a negative constant multiple of the angle, we get a pair of
spirals "upside-down." Illustration for Problem 6-4.

*clockwise. Negative angles go clockwise. Negative radii go in the opposite direc-
tion from the way you define the angle.*

Still Struggling

Look again at Fig. 6-2. If you ponder it for awhile, you might suspect that the
indicated equations aren't the only ones that can represent the lines shown. You
might ask, "If we allow r to range over all real numbers, can the line for $\theta = \pi/3$
also be represented by other equations such as $\theta = 4\pi/3$ or $\theta = -2\pi/3$? Can the
line for $\theta = 7\pi/8$ also be represented by $\theta = 15\pi/8$ or $\theta = -\pi/8$?" The answers
to these questions are both "Yes." When you see an equation of the form $\theta = a$
representing a straight line through the origin in polar coordinates, you can add
any integer multiple of π to the constant a, and you'll get another equation
whose graph shows up as the exact same line. In more formal terms, you can
represent the line $\theta = a$ through the origin in the form

$$\theta = k\pi + a$$

where k represents an integer constant and a represents a real-number constant.

PROBLEM 6-2

If each radial division (distance between concentric coordinate circles) represents precisely 1 unit in Fig. 6-4, what's the value of the constant a for the function as portrayed by the pair of spirals? What's the equation of the pair of spirals?

SOLUTION

Note that if $\theta = \pi$, then $r = 2$. You can solve for a by substituting this number pair in the general equation for a pair of spirals. Start with

$$r = a\theta$$

Plugging in the numbers $(\theta, r) = (\pi, 2)$, you get

$$2 = \pi a$$

which you can rearrange as

$$2/\pi = a$$

Now you know that $a = 2/\pi$ so the equation you seek is

$$r = (2/\pi)\theta$$

which you can also write as

$$r = 2\theta/\pi$$

PROBLEM 6-3

What's the polar equation of a straight line running through the origin and ascending at an angle of $\pi/4$ as you move to the right, with the restriction that $0 \le \theta < 2\pi$? If you draw this same line on a standard Cartesian xy coordinate grid instead of the polar plane, what equation does it represent?

SOLUTION

You can answer the first part of this problem with either of the equations

$$\theta = \pi/4$$

or

$$\theta = 5\pi/4$$

Keep in mind that the value of r can range over the entire set of real numbers —positive, negative, or zero.

First, look at the situation where $\theta = \pi/4$. When $r > 0$, you get a ray in the $\pi/4$ direction. When $r < 0$, you get a ray in the $5\pi/4$ direction. When $r = 0$, you get the origin point. The combination of these two rays and the origin forms the line running through the origin and ascending at an angle of $\pi/4$ as you move toward the right.

Now examine events with the equation $\theta = 5\pi/4$. When $r > 0$, you get a ray in the $5\pi/4$ direction. When $r < 0$, you get a ray in the $\pi/4$ direction. When $r = 0$, you get the origin point. The combination of the two rays and the origin forms the same line as you have in the first case.

To answer the second part of the problem, you can see that this line forms the graph of the equation $y = x$ in the Cartesian xy coordinate plane. If you can't envision the situation in your "mind's eye," draw graphs of the two polar equations (in polar coordinates, of course) and the Cartesian equivalent in the xy-plane. You'll get the same straight line in all three cases; it passes through the origin and runs off going "northeast" (upward and toward the right) and "southwest" (downward and toward the left).

PROBLEM 6-4

In Fig. 6-5, suppose that each radial increment represents π units. What's the value of the constant a in this case? What's the equation of the pair of spirals?

SOLUTION

Imagine a ray that points straight to the right along the reference axis labeled 0. As the ray rotates counterclockwise so that θ starts out at 0 and increases positively, the corresponding radius r starts out at 0 and increases negatively. These observations tell you that the constant a is negative. When the ray has turned through 1/2 rotation so that $\theta = \pi$, the radius of the solid spiral reaches the value $r = -2\pi$. (Don't get this confused with the apparent radius of $r = 4\pi$ on the solid spiral! The larger value is actually $r = -4\pi$, which you get when the ray has rotated through a complete circle so that $\theta = 2\pi$.) You can solve for a by substituting the number pair $(\theta, r) = (\pi, -2\pi)$ in the general spiral equation

$$r = a\theta$$

Plugging in the known values yields

$$-2\pi = a\pi$$

which solves to $a = -2$. Therefore, the equation of the pair of spirals is

$$r = -2\theta$$

Coordinate Transformations

It's easy to translate the coordinates of a point from the polar plane to the Cartesian plane, sort of like floating down a river from the source to the sea. It's more difficult to translate from Cartesian to polar form; that's like rowing up the same river from the sea to the source. As you read through this section, refer to Fig. 6-6, which shows a point in the polar grid superimposed on the Cartesian grid.

Polar to Cartesian

Consider a point (θ, r) in polar coordinates. We can convert these coordinates to Cartesian form (x, y) using the formulas

$$x = r \cos \theta$$

and

$$y = r \sin \theta$$

To envision how this conversion method works, imagine what happens when $r = 1$. The equation $r = 1$ in polar coordinates gives us a unit circle.

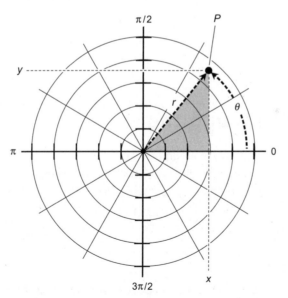

FIGURE 6-6 · A point plotted in both polar and Cartesian coordinates. Each radial division in the polar grid represents 1 unit. Each division on the x and y axes of the Cartesian grid also represents 1 unit. The shaded region portrays a right triangle x units wide, y units tall, and having a hypotenuse r units long.

In Chap. 3, we learned that for any point (x,y) on a unit circle in the Cartesian plane

$$x = \cos \theta$$

and

$$y = \sin \theta$$

Suppose that we double the radius of the circle, so its radius equals 2 units instead of 1 unit. Then we have the polar equation $r = 2$. The values of x and y in Cartesian coordinates both double, because when we double the length of the hypotenuse of a right triangle (such as the shaded region in Fig. 6-6), we also double the lengths of the other two sides. The new triangle is directly similar to the old one, meaning that its sides stay in the same ratio as we go around either triangle in the same direction. Therefore

$$x = 2 \cos \theta$$

and

$$y = 2 \sin \theta$$

This scheme works no matter how large or small we make the circle, as long as it stays centered at the origin. If $r = a$, where a represents an arbitrary positive real number, the new right triangle remains directly similar to triangle for $r = 1$, so

$$x = a \cos \theta$$

and

$$y = a \sin \theta$$

Still Struggling

If the radius has a negative real-number value, the foregoing formulas work just as well as they do when the radius is positive, even though the situation doesn't lend itself very well to the "mind's eye." For "extra credit," can you figure out why the formulas work for negative values of the radius?

PROBLEM 6-5

Consider the point $(\theta,r) = (\pi, 2)$ in polar coordinates. Determine the (x,y) representation of this point in Cartesian coordinates using the conversion formulas

$$x = r \cos \theta$$

and

$$y = r \sin \theta$$

SOLUTION

When we input π for the angle θ and 2 for the radius r, we get

$$x = 2 \cos \pi$$
$$= 2 \times (-1)$$
$$= -2$$

and

$$y = 2 \sin \pi$$
$$= 2 \times 0$$
$$= 0$$

We represent the polar point $(\theta, r) = (\pi, 2)$ in Cartesian coordinates as $(x, y) = (-2, 0)$.

Cartesian to Polar: The Radius

Figure 6-6 shows us that the radius r from the origin to the point $P = (x, y)$ equals the length of the hypotenuse of a right triangle (the shaded region) measuring x units wide and y units tall. Using the Pythagorean Theorem, we can write the formula for determining r in terms of x and y as

$$r = (x^2 + y^2)^{1/2}$$

That conversion goes easily enough! Now, let's work on the more difficult aspect of the conversion from Cartesian to polar coordinates: finding the polar angle for a generalized point in the Cartesian plane.

The Arctangent Function Revisited

Before we can find the polar direction angle for a point in Cartesian coordinates, let's remember the Arctangent function, which we discovered in Chap. 4. Figure 6-7 shows the graph of the principal branch of the Arctangent function (a duplicate of Fig. 4-13, repeated here for convenience). The Arctangent function "undoes" the work of the tangent function. Consider, for example, the fact that

$$\tan (\pi/4) = 1$$

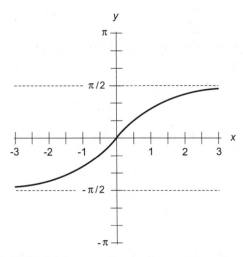

FIGURE 6-7 · Principal branch of the Arctangent function.

We can restate this equation using the Arctangent function to get

$$\text{Arctan}\,1 = \pi/4$$

For any real number u except odd-integer multiples of $\pi/2$ (for which the tangent function has no defined value), we can have complete confidence that

$$\text{Arctan}\,(\tan u) = u$$

Going the other way, for any real number v, we know that

$$\tan(\text{Arctan}\,v) = v$$

Cartesian to Polar: The Angle

We now have all the tools that we'll need to determine the polar angle θ for a point on the basis of its Cartesian coordinates x and y. We already know that

$$x = r\cos\theta$$

and

$$y = r\sin\theta$$

As long as $x \neq 0$, it follows that

$$
\begin{aligned}
y/x &= (r\sin\theta)/(r\cos\theta) \\
&= (r/r)\,(\sin\theta)/(\cos\theta) \\
&= (\sin\theta)/(\cos\theta) \\
&= \tan\theta
\end{aligned}
$$

which we can summarize and restate as

$$\tan \theta = y/x$$

If we take the Arctangent of both sides of this equation, we get

$$\text{Arctan}\,(\tan \theta) = \text{Arctan}\,(y/x)$$

which we can rewrite as

$$\theta = \text{Arctan}\,(y/x)$$

Suppose that the point $P = (x,y)$ lies in either the first quadrant or fourth quadrant of the Cartesian plane. In this case, we have

$$-\pi/2 < \theta < \pi/2$$

so we can directly use the conversion formula

$$\theta = \text{Arctan}\,(y/x)$$

If $P = (x,y)$ lies in the second or third quadrant of the Cartesian plane, then

$$\pi/2 < \theta < 3\pi/2$$

That's outside the range of the Arctangent function as graphed in Fig. 6-7, but we can remedy this situation if we subtract π from θ. That action brings θ into the allowed range (covered by the principal branch in the graph) but we don't change its tangent, because the tangent function repeats itself every π radians (half rotation). In this situation, we have

$$\theta - \pi = \text{Arctan}\,(y/x)$$

which we can restate as

$$\theta = \pi + \text{Arctan}\,(y/x)$$

Now, at last, we can derive specific formulas for θ in terms of x and y. Let's break the scenario down into all possible general locations for $P = (x,y)$, and see what we get for θ in each case.

P at the origin. If $x = 0$ and $y = 0$, then θ is theoretically undefined. However, let's assign θ a default value of 0 at the origin. This little bit of "cheating" allows us to fill the gap that would otherwise exist in our conversion scheme.

P on the $+x$ axis. If $x > 0$ and $y = 0$, then we're on the positive x axis. We can see from Fig. 6-6 that $\theta = 0$.

P in the first quadrant. If $x > 0$ and $y > 0$, then we're in the first quadrant of the Cartesian plane where θ is larger than 0 but less than $\pi/2$. We can therefore directly apply the conversion formula

$$\theta = \text{Arctan}\,(y/x)$$

P on the +y axis. If $x = 0$ and $y > 0$, then we're on the positive y axis. We can see from Fig. 6-6 that $\theta = \pi/2$.

P in the second quadrant. If $x < 0$ and $y > 0$, then we're in the second quadrant of the Cartesian plane where θ is larger than $\pi/2$ but less than π. In this case, we must apply the modified conversion formula

$$\theta = \pi + \text{Arctan}\,(y/x)$$

P on the −x axis. If $x < 0$ and $y = 0$, then we're on the negative x axis. We can see from Fig. 6-6 that $\theta = \pi$.

P in the third quadrant. If $x < 0$ and $y < 0$, then we're in the third quadrant of the Cartesian plane where θ is larger than π but less than $3\pi/2$, so we should apply the modified conversion formula

$$\theta = \pi + \text{Arctan}\,(y/x)$$

P on the −y axis. If $x = 0$ and $y < 0$, then we're on the negative y axis. We can see from Fig. 6-6 that $\theta = 3\pi/2$.

P in the fourth quadrant. If $x > 0$ and $y < 0$, then we're in the fourth quadrant of the Cartesian plane where θ is larger than $3\pi/2$ but smaller than 2π. That's the same thing as saying that $-\pi/2 < \theta < 0$. We'll get an angle in that range if we apply the original conversion formula

$$\theta = \text{Arctan}\,(y/x)$$

In polar coordinates, we should try to keep the angle positive but less than 2π. When we're in the fourth quadrant of the Cartesian plane, we can force the polar angle into the range $0 < \theta < 2\pi$ by adding a full rotation of 2π to the basic conversion formula, thereby getting

$$\theta = 2\pi + \text{Arctan}\,(y/x)$$

The foregoing nine cases account for all possible locations that a point P can have in a coordinate plane. Here's a summary of the conversion scheme that we've developed.

- $\theta = 0$ at the origin (by default)
- $\theta = 0$ on the $+x$ axis

- $\theta = \mathrm{Arctan}\,(y/x)$ in the first quadrant
- $\theta = \pi/2$ on the $+y$ axis
- $\theta = \pi + \mathrm{Arctan}\,(y/x)$ in the second quadrant
- $\theta = \pi$ on the $-x$ axis
- $\theta = \pi + \mathrm{Arctan}\,(y/x)$ in the third quadrant
- $\theta = 3\pi/2$ on the $-y$ axis
- $\theta = 2\pi + \mathrm{Arctan}\,(y/x)$ in the fourth quadrant

PROBLEM 6-6

Convert the Cartesian point $(-5,-12)$ to polar form. Assume that these coordinate values are exact.

SOLUTION

In this situation, we have the Cartesian coordinate values $x=-5$ and $y=-12$. When we plug these numbers into the Cartesian-to-polar conversion formula for the radius r, we get

$$r = [(-5)^2 + (-12)^2]^{1/2}$$
$$= (25 + 144)^{1/2}$$
$$= 169^{1/2}$$
$$= 13$$

Because both x and y are negative, our point lies in the third quadrant of the Cartesian plane. To find the angle in that case, we use the formula

$$\theta = \pi + \mathrm{Arctan}\,(y/x)$$

When we input the values $x=-5$ and $y=-12$, we get

$$\theta = \pi + \mathrm{Arctan}\,[(-12)/(-5)]$$
$$= \pi + \mathrm{Arctan}\,(12/5)$$

That's a theoretically exact, although arcane, answer. A calculator set to work in radians (not in degrees!) tells us that

$$\mathrm{Arctan}\,(12/5) = \mathrm{Arctan}\,2.4000$$
$$= 1.1760$$

rounded to four decimal places. If we let $\pi = 3.1416$, also rounded to four decimal places, we get

$$\theta = 3.1416 + 1.1760$$

$$= 4.3176$$

The polar equivalent of $(x,y) = (-5,-12)$ is therefore

$$(\theta,r) = (4.3176, 13.0000)$$

when we approximate the angle and the radius both to four decimal places.

Still Struggling

If the foregoing angle-conversion procedure baffles you, set it aside for awhile. Read it again tomorrow! You might want to "invent" some problems with points in all four quadrants of the Cartesian plane, and then use the formulas to convert the Cartesian coordinates to polar form. As you work out the arithmetic, you'll gain an understanding of how (and why) the formulas work.

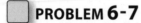 **PROBLEM 6-7**

Find the distance d in *radial units* between the following two points in polar coordinates, where we define a radial unit as the radius of a unit circle centered at the origin. Assume that the coordinate values are exact, and have the ordered pairs

$$P = (\pi, 3)$$

and

$$Q = (\pi/2, 4)$$

 **SOLUTION**

We can convert the polar coordinates of P and Q to Cartesian coordinates, and then employ the Cartesian distance formula to determine how far apart the two points lie. Let's call the Cartesian versions of the points

$$P = (x_p, y_p)$$

and

$$Q = (x_q, y_q)$$

For P, we have

$$x_p = 3 \cos \pi$$
$$= 3 \times (-1)$$
$$= -3$$

and

$$y_p = 3 \sin \pi$$
$$= 3 \times 0$$
$$= 0$$

The Cartesian coordinates of P are therefore $(x_p, y_p) = (-3, 0)$. For Q, we have

$$x_q = 4 \cos \pi/2$$
$$= 4 \times 0$$
$$= 0$$

and

$$y_q = 4 \sin \pi/2$$
$$= 4 \times 1$$
$$= 4$$

The Cartesian coordinates of Q are therefore $(x_q, y_q) = (0, 4)$. Using the Cartesian distance formula from Chap. 2, we obtain

$$d = [(x_p - x_q)^2 + (y_p - y_q)^2]^{1/2}$$
$$= [(-3 - 0)^2 + (0 - 4)^2]^{1/2}$$
$$= [(-3)^2 + (-4)^2]^{1/2}$$
$$= (9 + 16)^{1/2}$$
$$= 25^{1/2}$$
$$= 5$$

We've found that the points $P = (\pi, 3)$ and $Q = (\pi/2, 4)$ lie precisely 5 radial units apart in the polar coordinate plane.

PROBLEM 6-8

Figure 6-8 shows a line L and a circle C in polar coordinates. Line L passes through the origin. Every point on L lies equally far away from the horizontal

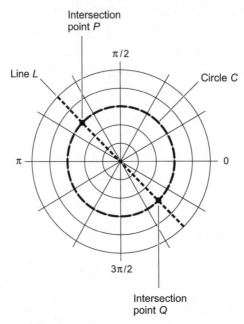

FIGURE 6-8 · Illustration for Problems 6-8 and 6-9.

axis (which runs from π to 0) and the vertical axis (which runs from $\pi/2$ to $3\pi/2$). Circle C has the origin as its center. Each radial division represents 1 unit. What's the polar equation representing L when we restrict the angle to positive values smaller than 2π? What's the polar equation representing C? Here's a hint: We can represent either equation in two ways.

SOLUTION

Line L runs *upward and to the left* from the origin at an angle halfway between the $\pi/2$ axis and the π axis. We represent that direction as the angle

$$\theta = 3\pi/4$$

That's a perfectly reasonable polar-coordinate equation of L. But we can also imagine that line L runs *downward and to the right* from the origin at an angle corresponding to

$$\theta = 7\pi/4$$

so this expression can also serve as the polar equation of L. Theoretically, we can add or subtract any integer multiple of π from $3\pi/4$ and get a valid equation for L. By convention, we should remain within the range of angles $0 \leq \theta < 2\pi$, so the above two equations are the preferred versions.

Circle C has the origin as its center. It has a radius of 3 units, as we can see by inspecting the graph and remembering that each radial division equals 1 unit. Therefore, we can represent C as the simple polar equation

$$r = 3$$

We can also consider the radius as −3 units, giving us the alternative equation

$$r = -3$$

PROBLEM 6-9

When we examine Fig. 6-8, we can see that L and C intersect at two points P and Q. What are the polar coordinates of P and Q, based on the information in Problem 6-8 and its solution? Here's a hint: We can represent both points in two ways.

 SOLUTION

Based on the solution to Problem 6-8, we can represent the intersection point at the upper left as either

$$P = (3\pi/4, 3)$$

or

$$P = (7\pi/4, -3)$$

We can represent the intersection point at the lower right as either

$$Q = (7\pi/4, 3)$$

or

$$Q = (3\pi/4, -3)$$

The more intuitive representations are the ones with the positive radii, which are

$$P = (3\pi/4, 3)$$

and

$$Q = (7\pi/4, 3)$$

The Navigator's Way

Navigators and military people use a coordinate plane similar to the one pre-ferred by mathematicians. The radius is called the *range*, expressed or measured in real-world units such as meters (m) or kilometers (km). The angle, or direc-tion, is called the *azimuth*, *heading*, or *bearing*, expressed in degrees clockwise from north. Figure 6-9 shows the layout for this system. Let's call it *navigator's polar coordinates* (NPC). We symbolize the azimuth as α (the lowercase italic Greek alpha), and we symbolize the range as r. We define the position of a point as an ordered pair (α, r).

What Is North?

There are two ways of defining "due north," or 0°. The more preferred standard uses *geographic north*, the direction in which you should travel if you want to take the shortest possible route over the earth to the north geographic pole, where our planet's rotational axis "emerges" from the surface. The less accurate standard uses *geomagnetic north*, the direction indicated by the needle in a mag-netic compass.

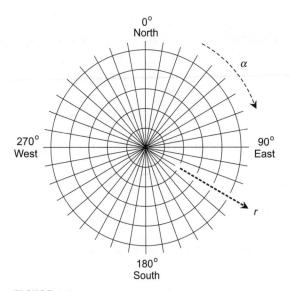

FIGURE 6-9 · The navigator's polar coordinate (NPC) plane. We express the azimuth α in degrees clockwise from north. We express the range r as a straight-line distance from our position (the origin).

For most locations on the earth's surface, a difference exists between geographic north and geomagnetic north. Astronomers call this angular difference, measured in degrees, the *magnetic declination*. (Beware! Magnetic declination has nothing to do with the *celestial declination* coordinate that astronomers use to denote objects in the sky!) Navigators in olden times had to know the magnetic declination for their location when they couldn't use the stars to determine geographic north.

TIP *Nowadays, we have electronic navigation schemes such as the* **Global Positioning System** *(GPS) that offer far greater accuracy than any magnetic compass ever did—as long as those sophisticated systems keep working.*

Strict Constraints

When we use NPC, we must never allow the range to attain negative values. No navigator talks about having gone −20 km on a heading of 270°, for example, when they mean to say that they traveled 20 km on a heading of 90°. When working out certain problems in NPC, the result might end up containing a negative range. In that case, multiply r by −1 to make it positive, and then increase or decrease α by 180° to get a nonnegative angle less than 360°.

The azimuth, bearing, or heading must also conform to certain values. The smallest possible value of α is 0° (representing geographic north). As you turn clockwise as seen from above, the values of α increase through 90° (east), 180° (south), 270° (west), and ultimately approach, but never reach, 360° (north again). We therefore have these restrictions on the ordered pair (α, r):

$$0° \le \alpha < 360°$$

and

$$r \ge 0$$

Polar-to-Polar Conversion

Once in awhile, you'll want to convert from the previously defined system of *mathematician's polar coordinates* (MPC) to NPC or vice-versa. When you carry out such a *coordinate transformation*, a point's radius remains the same, but its direction angle usually changes.

If you know the direction θ_0 of a point in MPC and you want to find the equivalent azimuth α_0 in NPC, multiply the radian measure by 180/π to get degrees. Then use one or the other of the following formulas, depending on the value of θ_0 in degrees:

$$\alpha_0 = 90° - \theta_0 \text{ if } 0° \le \theta_0 \le 90°$$

or

$$\alpha_0 = 450° - \theta_0 \text{ if } 90° < \theta_0 < 360°$$

If you know the azimuth α_0 of a distant point in NPC and you want to find the equivalent direction θ_0 in MPC, multiply the degree measure by $\pi/180$ to get radians. Then use one or the other of the following formulas, depending on the value of α_0 in radians:

$$\theta_0 = \pi/2 - \alpha_0 \text{ if } 0 \le \alpha_0 \le \pi/2$$

or

$$\theta_0 = 5\pi/2 - \alpha_0 \text{ if } \pi/2 < \alpha_0 < 2\pi$$

Navigator's Polar versus Cartesian

Suppose that you want to convert from NPC to Cartesian coordinates. The conversion formulas for translating the coordinates for a point (α_0, r_0) in NPC to a point (x_0, y_0) in the Cartesian plane are

$$x_0 = r_0 \sin \alpha_0$$

and

$$y_0 = r_0 \cos \alpha_0$$

These formulas resemble the ones for conversion from MPC to Cartesian coordinates, except that the sine and cosine functions play the opposite roles.

In order to convert the coordinates of a point (x_0, y_0) in Cartesian coordinates to a point (α_0, r_0) in NPC, you can convert to MPC first, and then convert to NPC.

PROBLEM 6-10

Imagine that an NPC radar set reveals the presence of a stationary airborne object at azimuth 300° and range 40 km. If we say that a kilometer equals a "unit," what are the coordinates (θ_0, r_0) of this object in MPC?

 SOLUTION

We're given the coordinates $(\alpha_0, r_0) = (300°, 40)$. The value of r_0, the radius, equals the range, in this case 40 units. Multiplying degrees by $\pi/180$, we convert the angle from 300° to get a radian measure of $5\pi/3$. We see that α_0 exceeds $\pi/2$ but remains less than 2π, so we should use the formula

$$\theta_0 = 5\pi/2 - \alpha_0$$

Inputting $5\pi/3$ for α_0, we get

$$\theta_0 = 5\pi/2 - 5\pi/3$$
$$= 15\pi/6 - 10\pi/6$$
$$= 5\pi/6$$

Therefore, $(\theta_0, r_0) = (5\pi/6, 40)$.

PROBLEM 6 -11

Imagine that you're wandering through the wilderness on an archeological expedition. You unearth a flat stone tablet with a treasure map chiseled into its face. The map says "You are here" next to an X, and then says, "Go north 40 paces and then west 30 paces." Let west represent the negative x axis of a Cartesian coordinate system; let east represent the positive x axis; let south represent the negative y axis; let north represent the positive y axis. Let one "pace" represent one "unit" of radius, and also one "unit" in the Cartesian system. If you're naïve enough to look for the treasure and lazy enough to insist on walking in a straight line to reach it, how many paces should you travel, and in what direction, in NPC? Determine your answer to the nearest degree, and to the nearest pace.

SOLUTION

First, you must determine the ordered pair in Cartesian coordinates that corresponds to the imagined treasure site. Consider the origin as the spot where you unearthed the tablet. If you let (x_0, y_0) represent the point where you hope to find the treasure, then 40 paces north means $y_0 = 40$, and 30 paces west means $x_0 = -30$. You therefore have

$$(x_0, y_0) = (-30, 40)$$

Because $x_0 < 0$ and $y_0 > 0$, you calculate the MPC direction angle θ_0 as

$$\theta_0 = \pi + \text{Arctan}\,(y_0/x_0)$$
$$= \pi + \text{Arctan}\,[40/(-30)]$$
$$= \pi + \text{Arctan}\,(-4/3)$$

You'll want to use degrees in your final expression, so you might as well rewrite the above equation in terms of degrees right now, and solve it as

$$\theta_0 = 180° + \text{Arctan}\,(-4/3)$$
$$= 180° + (-53°)$$
$$= 127°$$

To convert the MPC angle to the NPC angle in this case, you subtract the MPC angle from 450°. In this case you get

$$\alpha_0 = 450° - 127°$$
$$= 323°$$

The radius r_0 is the length of a right angle's hypotenuse. It works out as

$$r_0 = (x_0{}^2 + y_0{}^2)^{1/2}$$
$$= [(-30)^2 + 40^2]^{1/2}$$
$$= 50$$

The radius r_0 in NPC always equals the radius r_0 in MPC, so you know that your NPC coordinates are

$$(\alpha_0, r_0) = (323°, 50)$$

Walk 50 paces at a heading of 323° (approximately north by northwest). Then dig if you dare. If you find the treasure, congratulations! If you find nothing other than dirt and stones, then the joke's on you.

QUIZ

Refer to the text in this chapter if necessary. A good score is eight correct. Answers are in the back of the book.

1. What are the Cartesian *xy*-plane coordinates of the mathematician's polar coordinate (MPC) point $(5\pi/4, 2^{1/2})$? Remember that the 1/2 power means the positive square root.

    A. (1,1)
    B. (−1,1)
    C. (−1,−1)
    D. (1,−1)

2. What are the MPC values of the Cartesian *xy*-plane point (4,−3), if we want to specify the angle in degrees rather than in radians?

    A. $(37°, 4)$
    B. $(127°, 12^{1/2})$
    C. $(300°, 12)$
    D. $(323°, 5)$

3. What are the navigator's polar coordinate (NPC) values of the Cartesian *xy*-plane point (4,−3)?

    A. $(37°, 5)$
    B. $(270°, 12)$
    C. $(127°, 5)$
    D. $(323°, 12^{1/2})$

4. Figure 6-10 shows two straight lines and a circle in the MPC plane. What's the equation of line *L*? Assume that each radial increment (distance between any pair of concentric circles) represents π units.

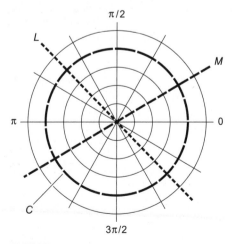

FIGURE 6-10 • Illustration for Quiz Questions 4 through 6.

A. $r = 3\pi/4$
B. $\theta = 3\pi/4$
C. $\theta = 4$
D. $r = \pi/6$

5. What's the MPC equation of line M in Fig. 6-10? Assume that each radial increment (distance between any pair of concentric circles) represents π units.

A. $\theta = \pi/6$
B. $r = \pi$
C. $\theta = 3\pi/4$
D. $r = \pi/6$

6. What's the MPC equation of circle C in Fig. 6-10? Assume that each radial increment (distance between any pair of adjacent concentric circles) represents π units.

A. $r = 4\pi$
B. $r = \pi/6$
C. $\theta = \pi/4$
D. $\theta = 7\pi/4$

7. Imagine that a radar set with an NPC display indicates the presence of a stationary airborne object at an azimuth (compass) bearing of 45° and a range of 28.3 km. If we say that a kilometer equals a "unit," what are the NPC coordinates of this object?

A. $(135°, 28.3)$
B. $(45°, 28.3)$
C. $(225°, 20)$
D. $(315°, 20)$

8. If we say that a kilometer equals a "unit" along both axes in a Cartesian plane, what are the Cartesian xy-plane coordinates of the object described in Question 7, rounded off to the nearest whole-integer units?

A. $(20, -20)$
B. $(-20, -20)$
C. $(-20, 20)$
D. $(20, 20)$

9. Which of the following coordinates can't "legally" have a negative value?

A. The x coordinate in a Cartesian xy-plane
B. The y coordinate in a Cartesian xy-plane
C. The radius in an MPC plane
D. The range in an NPC plane

10. Someone assures us that we can represent an object in the MPC plane with the simple equation $r = -16$. What's the equation of this object in Cartesian coordinates?

A. $x^2 + y^2 = 4$
B. $x^2 + y^2 = 16$
C. $x^2 + y^2 = 256$
D. None exists, because we can't "legally" have a negative radius coordinate value in MPC.

Three-Space and Vectors

Let's leave two-dimensional (2D) planes and venture into three-dimensional (3D) space, also called *three-space*. We gain freedom, but we encounter complexity! Then we'll learn about *vectors*, which express physical quantities such as displacement, velocity, acceleration, and force.

CHAPTER OBJECTIVES

In this chapter, you will

- Use latitude and longitude coordinates to define points on the earth's surface.
- Use latitude and longitude coordinates to define directions in space.
- Arrange a set of rectangular coordinates to define points in space.
- Use cylindrical and spherical coordinates to define points in space.
- Acquaint yourself with vector arithmetic in two and three dimensions.

Spatial Coordinates

Mathematicians, scientists, and engineers employ several different coordinate systems to define points and create graphs in 3D space, also known as *three-space*.

Latitude and Longitude

Latitude and *longitude* angles uniquely define the positions of points on the surface of a sphere or in the sky. Figure 7-1A illustrates the basic arrangement for *terrestrial coordinates*, the most common scheme for pinpointing locations on the earth's surface. In this system, the origin, where all axes intersect, lies at the center of the earth. The *polar axis* connects two specified points at *antipodes*, or points directly opposite each other, on the earth's surface. We assign these values latitude coordinates of $\theta = 90°$ (the north geographic pole) and $\theta = -90°$ (the south geographic pole). The earth rotates on the polar axis. The *equatorial axis* runs outward from the origin at a right angle to the polar axis, passing through an agreed-on surface point midway between the geographic poles. We assign the longitude coordinate $\phi = 0°$ to the equatorial axis.

We can measure or express latitude θ either northward (positively) or southward (negatively) relative to the plane of the earth's equator. We can measure or express longitude ϕ going around the planet either counterclockwise (positively)

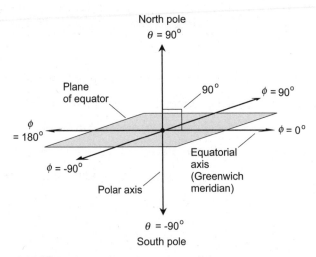

FIGURE 7-1A · Latitude and longitude angles for locating points on the earth's surface.

or clockwise (negatively) relative to the equatorial axis as viewed from above the north pole. To avoid ambiguity in the coordinate system, we restrict latitude values to

$$-90° \leq \theta \leq +90°$$

and longitude values to

$$-180° < \phi \leq +180°$$

On the earth's surface, a special longitude half-circle connects the geographic poles and also passes through Greenwich, England. Scientists and navigators call this half-circle the *Greenwich meridian* or the *prime meridian*, defining all longitude angles with respect to it. The equatorial axis passes through the point where the Greenwich meridian, defined as longitude $\phi = 0°$, crosses the equator.

TIP *In ordinary mathematics, we rarely use the plus sign in front of positive numbers because the sign would be redundant. In the case of latitude or longitude numbers, however, we'll often see it included (as in the above examples). As long as we know the context, we can get away without including plus signs in front of positive latitude or longitude numbers (as in Figs. 7-1A and 7-1B, for example).*

TIP *Besides allowing us to uniquely locate places on the earth's surface, latitude and longitude angles can translate into positions in the sky as we see it from the earth's surface. These places in the sky aren't actual points, but instead correspond to rays pointing out from our eyes indefinitely into space, defining specific directions or orientations. These rays revolve in the cosmos, completing one revolution every 24 hours with respect to the sun, and one revolution every 23 hours and 56 minutes (approximately) relative to the distant stars and galaxies. In this manner, we can define a system of coordinates for points as we see them in the sky.*

Celestial Latitude and Longitude

Celestial latitude and *celestial longitude* extend the earth's latitude and longitude angles into the heavens. The same set of coordinates that we use for geographic latitude and longitude applies to the celestial system as well. An object whose celestial latitude and longitude coordinates are (θ,ϕ) appears at the *zenith* in the sky (directly overhead) from the point on the earth's surface with latitude and longitude coordinates (θ,ϕ).

Declination and Right Ascension

Declination and *right ascension* define the positions of objects in the sky relative to the stars, rather than relative to the earth. Figure 7-1B applies to this system, whose "grid in the sky" remains fixed as the earth rotates on its axis.

We measure declination (θ) northward and southward with respect to the *celestial equator,* an imaginary circle in the heavens that runs from the eastern horizon to the western horizon, and passes through the zenith for anyone standing on the earth's surface at the equator. Declination and celestial latitude are identical. The *north celestial pole* (declination 90°) appears at the zenith for a person standing on the earth's surface at the north geographic pole. The *south celestial pole* (declination –90°) appears at the zenith for a person standing on the earth's surface at the south geographic pole.

We measure right ascension (ϕ) going eastward from the *vernal equinox,* the location of the sun in the heavens at the moment spring begins in the northern hemisphere (usually on March 20 or 21). Instead of expressing right ascension angles in degrees or radians, astronomers use units called *hours, minutes,* and *seconds* based on 24 hours in a complete circle, corresponding to the 24 hours in a solar day. Therefore, one hour (1 h) of right ascension equals 15° of arc, 6 h = 90°, 12 h = 180°, and 18 h = 270°. Minutes and seconds of right ascension differ from the arc minutes and arc seconds used by mathematicians and engineers, about which you learned in Chap. 1. One minute (1 m) of celestial right

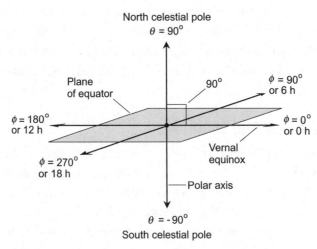

FIGURE 7-1B · Declination and right ascension angles for locating points in the sky.

ascension equals 1/60 of an hour or 1/4 of an angular degree, and one second (1 s) of celestial right ascension equals 1/60 of 1 m or 1/240 of an angular degree.

Rectangular Three-Space

When we extend rectangular coordinates into three dimensions, we get *rectangular three-space*, also called *xyz-space* (Fig. 7-2). We usually plot the two independent-variable values along the x and y axes, and the dependent-variable values along the z axis.

In rectangular three-space, each axis runs perpendicular to the other two. They all intersect at the *origin*, normally corresponding to the coordinate values $x = 0$, $y = 0$, and $z = 0$. We represent individual points in rectangular three-space as *ordered triples* of the form (x,y,z).

The scales in rectangular three-space are all linear. In other words, along any given individual axis, equal distances represent equal changes in value. But the divisions or increments (the spaces between hash marks) on different axes don't have to represent the same amounts of value change. For example, we might construct the x axis with one unit per division, the y axis with two units per division, and the z axis with five units per division.

When we lay out all three axes in a 3D rectangular coordinate system using increments of the same size (representing the same amounts of value change

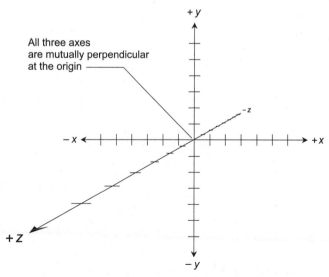

FIGURE 7-2 · Rectangular three-space, also called *xyz*-space.

per increment), then we have a specialized system of rectangular coordinates called *Cartesian three-space* or *Cartesian xyz-space*. In this system, we get an "undistorted" view of mathematical 3D space.

TIP *Normally, you shouldn't include spaces after the commas when denoting an ordered triple. Instead, you should write the whole expression as a single "run-on sequence." However, you can violate this rule if the variable expressions in the ordered triple grow so complicated or lengthy that you have trouble seeing the commas without allowing for a little bit of space on either side.*

Mathematician's Cylindrical Coordinates

Let's examine two similar systems of *cylindrical coordinates* for specifying the positions of points in 3D space. These systems arise when we move polar coordinate planes straight up and down so that the radius circles trace out *cylinders* in space.

Figure 7-3A shows a set of cylindrical coordinates that starts with rectangular xy-space superimposed on a mathematician's polar coordinate (MPC) plane. Then we define the z axis vertically through the origin of that plane. We define a rotational angle θ in the xy-plane, expressed in degrees or radians (usually radians) counterclockwise from the positive x axis, which forms the *reference axis*.

Imagine some point P in space, and consider its projection P' onto the xy-plane such that line segment PP' runs vertically, parallel to the z axis. We can uniquely

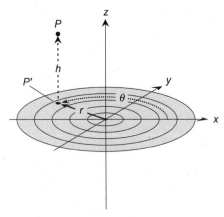

FIGURE 7-3A • Mathematician's cylindrical coordinates for defining points in three-space.

define the position of P as an ordered triple (θ, r, h) in which θ represents the angle measured counterclockwise from the x axis to P' in the xy-plane, r represents the distance or radius from the origin to P' in the xy-plane, and h represents the distance (altitude or height) of P above the xy-plane.

Navigator's Cylindrical Coordinates

In the system shown in Fig. 7-3B, we start with a rectangular xy-plane superimposed on a navigator's polar coordinate (NPC) plane that corresponds to the surface of the earth in the vicinity of the origin. The z axis runs straight up (positive z values) and down (negative z values), passing through the origin of the xy-plane. We define the angle θ in the xy-plane using degrees (never radians) going *clockwise* from the positive y axis, which corresponds to geographic north.

Given a point P in space, consider its projection P' onto the xy-plane, such that line segment PP' runs vertically, parallel to the z axis. We can uniquely define the position of P as an ordered triple (θ, r, h) in which θ represents the angle measured clockwise from geographic north to P', r represents the distance or radius from the origin to P' in the xy-plane, and h represents the distance (altitude or height) of P above the xy-plane.

Spherical Coordinates

Figure 7-4 shows three systems of *spherical coordinates* for defining points in space. Astronomers and aerospace scientists prefer the first two (A and B), while navigators and surveyors prefer the third one (C).

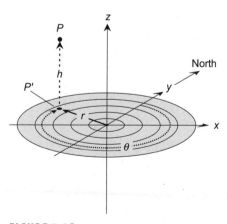

FIGURE 7-3B • Navigator's cylindrical coordinates for defining points in three-space.

In the scheme of Fig. 7-4A, we define the location of a point P as an ordered triple (θ,ϕ,r) such that θ represents the declination of P, ϕ represents the right ascension of P, and r represents the radius (also called the *range*) from the origin to P. In this example, we usually express the angles in degrees (except in the case of the astronomer's version of right ascension, which is expressed in hours, minutes, and seconds as defined earlier in this chapter). Alternatively, we can express the angles in radians. This system maintains a fixed orientation relative to the distant stars.

Instead of declination and right ascension, the variables θ and ϕ can represent celestial latitude and celestial longitude respectively, as shown in Fig. 7-4B. This system stays fixed relative to the earth, rather than relative to the background of distant stars. Therefore, the coordinates of all celestial objects constantly move as the earth rotates on its axis. Specialized satellites, known as *geostationary satellites* because their orbits carry them around the earth at the same rate as the earth rotates, remain fixed with respect to the system of Fig. 7-4B.

We have a third alternative for spherical coordinates. The angle θ can represent the *elevation* (angle above the horizon) and ϕ can represent the *azimuth* (bearing or heading) as measured clockwise from geographic north. In this case, the reference plane corresponds to the horizon at our location. The elevation can cover the span of values between, and including, $-90°$ (the *nadir*, or the point

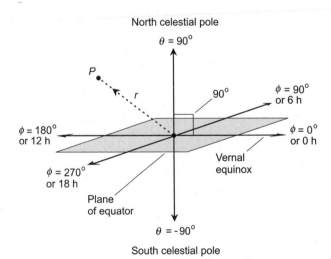

FIGURE 7-4A • Spherical coordinates for defining points in three-space, where the angles θ and ϕ represent declination and right ascension, and r represents radius or range.

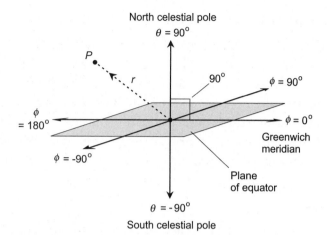

FIGURE 7-4B • Spherical coordinates for defining points in three-space, where the angles θ and ϕ represent latitude and longitude, and r represents radius or range.

directly underfoot) and 90° (the zenith). Figure 7-4C illustrates this spherical coordinate methodology.

In a variant of the system of Fig. 7-4C, we express the angle θ with respect to the zenith (or the positive z axis), rather than relative to the plane of the horizon. In that case, the range for the elevation angle is $0° \le \theta \le 180°$ (or $0° \le \theta \le \pi$). Mathematicians sometimes use this spherical-coordinate arrangement.

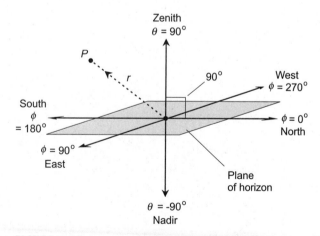

FIGURE 7-4C • Spherical coordinates for defining points in three-space, where the angles θ and ϕ represent elevation (angle above the horizon) and azimuth (also called bearing or heading), and r represents radius or range.

PROBLEM 7-1

Imagine that you fly a kite above a perfectly flat, level field. The wind blows out of the east-southeast, or azimuth 120°. Therefore, the kite flies in a west-north-westerly direction, at azimuth 300°. Suppose that the kite hovers at an elevation angle of 50° above the horizon, and the kite line measures 100 meters long. Imagine that the sun is exactly overhead so that the kite's shadow falls directly beneath it. How far from you (at the origin) does the kite's shadow lie? How high does the kite hover above the ground, as measured along a line going straight up and down? Express your answers to the nearest meter.

SOLUTION

You can work in navigator's cylindrical coordinates to solve this problem. You must consider the length of the kite line (100 meters) and the angle at which the kite flies (50°). Figure 7-5 shows the scenario. Let r represent the distance of the shadow from you, as expressed in meters. Let h represent the altitude of the kite above the ground, also in meters, measured along a vertical line.

First, find the ratio of the kite's altitude to the length of its line. The line segment whose length equals h, the line segment whose length equals r, and the kite line form a right triangle with the hypotenuse corresponding to the kite line. (Assume that the kite line doesn't sag, so it's perfectly straight.) From basic circular trigonometry, you know that

$$\sin 50° = h/100$$

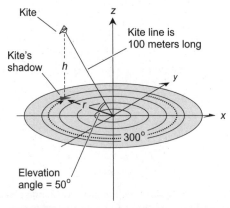

FIGURE 7-5 • Illustration for Problem 7-1.

Using a calculator, you can derive *h* as

$$\sin 50° = 0.766$$
$$= h/100$$

This equation solves to

$$h = 77 \text{ meters (rounded to the nearest meter)}$$

You also know, from basic circular trigonometry, the fact that

$$\cos 50° = r/100$$

Using a calculator, you can derive *r* as

$$\cos 50° = 0.643$$
$$= r/100$$

which solves to

$$r = 64 \text{ meters (rounded to the nearest meter)}$$

Still Struggling

In the foregoing situation, the wind direction doesn't make any difference. However, if the sun weren't directly overhead, the wind direction would affect your answer to this problem. The "nonzenith" solar position would render the calculation process a lot more complicated! If you like major challenges, try a variant of Problem 7-1 in which the sun shines from the southern sky (azimuth 180°), and its rays strike the earth's surface at an angle of 35° with respect to the horizontal.

Vectors in the Cartesian Plane

Mathematicians and scientists define a vector as a quantity with two independent properties called *magnitude* and *direction*. We can use vectors to represent physical variables such as displacement, velocity, and acceleration. Let's denote vectors as boldface letters of the alphabet. In the xy-plane, for example, we can illustrate two vectors **a** and **b** as rays (half-lines) from the origin $(0,0)$ to points (x_a, y_a) and (x_b, y_b), as shown in Fig. 7-6.

Magnitude of a Vector

We can calculate the magnitude (or length) of a vector **a**, written $|a|$ or a, in the Cartesian xy-plane with the distance formula

$$|a| = (x_a^2 + y_a^2)^{1/2}$$

Direction of a Vector

We define the direction (or orientation) of the vector **a**, written dir **a**, as the angle θ_a that the vector subtends as we rotate counterclockwise from the positive x axis. We can find this angle using the same scheme that we used for coordinate conversion in Chap. 6, modified for vectors. For angles in degrees:

- If $x_a > 0$ and $y_a = 0$, then $\theta_a = 0°$
- If $x_a > 0$ and $y_a > 0$, then $\theta_a = \text{Arctan}(y_a/x_a)$
- If $x_a = 0$ and $y_a > 0$, then $\theta_a = 90°$
- If $x_a < 0$ and $y_a > 0$, then $\theta_a = 180° + \text{Arctan}(y_a/x_a)$
- If $x_a < 0$ and $y_a = 0$, then $\theta_a = 180°$
- If $x_a < 0$ and $y_a < 0$, then $\theta_a = 180° + \text{Arctan}(y_a/x_a)$

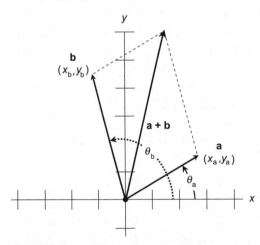

FIGURE 7-6 · Two vectors **a** and **b** in the Cartesian plane. Their sum runs along the diagonal of a parallelogram.

- If $x_a = 0$ and $y_a < 0$, then $\theta_a = 270°$
- If $x_a > 0$ and $y_a < 0$, then $\theta_a = 360° + \text{Arctan}\,(y_a / x_a)$

For angles in radians:

- If $x_a > 0$ and $y_a = 0$, then $\theta_a = 0$
- If $x_a > 0$ and $y_a > 0$, then $\theta_a = \text{Arctan}\,(y_a / x_a)$
- If $x_a = 0$ and $y_a > 0$, then $\theta_a = \pi/2$
- If $x_a < 0$ and $y_a > 0$, then $\theta_a = \pi + \text{Arctan}\,(y_a / x_a)$
- If $x_a < 0$ and $y_a = 0$, then $\theta_a = \pi$
- If $x_a < 0$ and $y_a < 0$, then $\theta_a = \pi + \text{Arctan}\,(y_a / x_a)$
- If $x_a = 0$ and $y_a < 0$, then $\theta_a = 3\pi/2$
- If $x_a > 0$ and $y_a < 0$, then $\theta_a = 2\pi + \text{Arctan}\,(y_a / x_a)$

Sum of Two Vectors

In the Cartesian plane, we can calculate the sum of two vectors **a** and **b**, where $\mathbf{a} = (x_a, y_a)$ and $\mathbf{b} = (x_b, y_b)$, with the formula

$$\mathbf{a} + \mathbf{b} = [(x_a + x_b), (y_a + y_b)]$$

We can express this sum in geometric form by constructing a parallelogram with the vectors **a** and **b** as adjacent sides. In that case, the vector **a** + **b** runs along the diagonal of the parallelogram, as shown in Fig. 7-6.

Multiplying a Vector by a Scalar

When we want to multiply a vector by a *scalar* (a "plain old real number"), we multiply both the x and y components of the vector by that scalar, getting a new vector. This operation obeys the *commutative principle*, meaning that it doesn't matter whether the scalar comes before or after the vector in the product. If we have a vector $\mathbf{a} = (x_a, y_a)$ and a scalar k, then

$$k\mathbf{a} = (kx_a, ky_a)$$

and

$$\mathbf{a}k = (x_a k, y_a k)$$
$$= (kx_a, ky_a)$$
$$= k\mathbf{a}$$

Dot Product

Consider two vectors called **a** and **b** in the Cartesian xy-plane with the coordinates

$$\mathbf{a} = (x_a, y_a)$$

and

$$\mathbf{b} = (x_b, y_b)$$

We define the *dot product*, also called the *scalar product* and written **a** • **b** (read out loud "a dot b"), as the real number produced by the formula

$$\mathbf{a} \bullet \mathbf{b} = x_a x_b + y_a y_b$$

 **PROBLEM 7-2**

What's the sum of the two vectors **a** = (3,−5) and **b** = (2,6) in the Cartesian plane?

 SOLUTION

We add the x and y components together independently to get

$$\mathbf{a} + \mathbf{b} = [(3 + 2), (−5 + 6)]$$
$$= (5, 1)$$

 **PROBLEM 7-3**

What's the dot product of the two vectors **a** = (3,−5) and **b** = (2,6)?

SOLUTION

We input the values into the formula for the dot product and do the arithmetic, getting

$$\mathbf{a} \bullet \mathbf{b} = (3 \times 2) + (−5 \times 6)$$
$$= 6 + (−30)$$
$$= −24$$

 **PROBLEM 7-4**

What happens if we reverse the order of the dot product of two vectors?

 **SOLUTION**

We get the same result regardless of the order in which we "dot-multiply" the vectors. Therefore, the dot product constitutes a commutative operation. We can prove this fact in the general case using the formula given above. Consider two vectors in **a** and **b** in the Cartesian xy-plane, such that

$$\mathbf{a} = (x_a, y_a)$$

and

$$\mathbf{b} = (x_b, y_b)$$

We calculate the dot product of **a** and **b**, in that order, as

$$\mathbf{a} \bullet \mathbf{b} = x_a x_b + y_a y_b$$

We calculate the dot product **b** • **a**, in that order, as

$$\mathbf{b} \bullet \mathbf{a} = x_b x_a + y_b y_a$$

In our prealgebra courses, we learned that "plain old real-number multiplication" obeys the commutative law. Therefore, we can rewrite the formula for **b** • **a** as

$$\mathbf{b} \bullet \mathbf{a} = x_a x_b + y_a y_b$$

We already know that $x_a x_b + y_a y_b$ represents the arithmetic expansion of **a** • **b**. Therefore, for any two vectors **a** and **b**, we can conclude that

$$\mathbf{a} \bullet \mathbf{b} = \mathbf{b} \bullet \mathbf{a}$$

Vectors in the MPC Plane

In mathematician's polar coordinates (MPC), we can denote vectors **a** and **b** as rays from the origin $(0,0)$ to points (θ_a, r_a) and (θ_b, r_b), as shown in Fig. 7-7.

Magnitude and Direction of a Vector

We can express the magnitude and direction of vector $\mathbf{a} = (\theta_a, r_a)$ in MPC simply and directly as

$$|\mathbf{a}| = r_a$$

and

$$\text{dir } \mathbf{a} = \theta_a$$

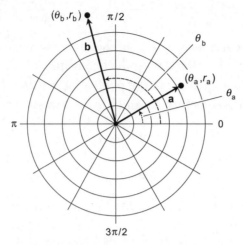

FIGURE 7-7 · Two vectors **a** and **b** in the mathematician's polar coordinate (MPC) plane. The angles are θ_a and θ_b. We express the angles in radians. The radii are r_a and r_b.

By convention, we hold to the following restrictions:

$$0° \le \theta_a < 360°$$
$$\text{for } \theta_a \text{ in degrees}$$

and

$$0 \le \theta_a < 2\pi$$
$$\text{for } \theta_a \text{ in radians}$$

We must always keep the radius nonnegative. We express that constraint in mathematical terms as

$$r_a \ge 0$$

Sum of Two Vectors

When we want to determine the sum of two vectors **a** and **b** in MPC, we can convert those vectors into their Cartesian xy-plane equivalents, add them according to the formula for vector addition in the xy-plane, and then convert the resultant vector back to MPC. To convert vector **a** from MPC to rectangular coordinates, we calculate

$$x_a = r_a \cos \theta_a$$

and

$$y_a = r_a \sin \theta_a$$

To convert vector a from rectangular coordinates to MPC, we can use the formulas we learned a little while ago for the direction angle. Let's restate them now for clarity! For angles in degrees:

- If $x_a > 0$ and $y_a = 0$, then $\theta_a = 0°$
- If $x_a > 0$ and $y_a > 0$, then $\theta_a = \text{Arctan}(y_a/x_a)$
- If $x_a = 0$ and $y_a > 0$, then $\theta_a = 90°$
- If $x_a < 0$ and $y_a > 0$, then $\theta_a = 180° + \text{Arctan}(y_a/x_a)$
- If $x_a < 0$ and $y_a = 0$, then $\theta_a = 180°$
- If $x_a < 0$ and $y_a < 0$, then $\theta_a = 180° + \text{Arctan}(y_a/x_a)$
- If $x_a = 0$ and $y_a < 0$, then $\theta_a = 270°$
- If $x_a > 0$ and $y_a < 0$, then $\theta_a = 360° + \text{Arctan}(y_a/x_a)$

For angles in radians:

- If $x_a > 0$ and $y_a = 0$, then $\theta_a = 0$
- If $x_a > 0$ and $y_a > 0$, then $\theta_a = \text{Arctan}(y_a/x_a)$
- If $x_a = 0$ and $y_a > 0$, then $\theta_a = \pi/2$
- If $x_a < 0$ and $y_a > 0$, then $\theta_a = \pi + \text{Arctan}(y_a/x_a)$
- If $x_a < 0$ and $y_a = 0$, then $\theta_a = \pi$
- If $x_a < 0$ and $y_a < 0$, then $\theta_a = \pi + \text{Arctan}(y_a/x_a)$
- If $x_a = 0$ and $y_a < 0$, then $\theta_a = 3\pi/2$
- If $x_a > 0$ and $y_a < 0$, then $\theta_a = 2\pi + \text{Arctan}(y_a/x_a)$

Regardless of the values of x_a and y_a, we can calculate the MPC radius r_a of a Cartesian-plane vector $\mathbf{a} = (x_a, y_a)$ as

$$r_a = (x_a^2 + y_a^2)^{1/2}$$

Multiplying a Vector By a Scalar

Let's define an MPC vector **a** with the coordinates (θ, r), as shown in Fig. 7-8. Suppose that we multiply **a** by a positive real scalar k. We have

$$k\mathbf{a} = (\theta, kr)$$

If we multiply **a** by a negative real scalar $-k$, then we get

$$-k\mathbf{a} = [(\theta \pm 180°), kr]$$

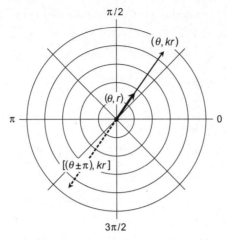

FIGURE 7-8 · Multiplication of an MPC-plane vector **a** by a positive real scalar k, and by a negative real scalar $-k$. We express the angles in radians.

for θ in degrees. For θ in radians, we have

$$-k\mathbf{a} = [(\theta \pm \pi), kr]$$

Still Struggling

The addition or subtraction of a half rotation (180° or π rad) to a vector's direction angle reverses the direction of that vector. Whether we should add or subtract a half rotation depends on the value of the result. We should always end up with a nonnegative direction angle representing less than a full rotation (360° or 2π rad).

Dot Product of Two Vectors

In the MPC plane, let r_a represent the radius of vector **a**, and let r_b represent the radius of vector **b**. We can calculate the dot product of **a** and **b** as

$$\mathbf{a} \bullet \mathbf{b} = |\mathbf{a}|\,|\mathbf{b}| \cos (\theta_b - \theta_a)$$
$$= r_a r_b \cos (\theta_b - \theta_a)$$

PROBLEM 7-5

Consider the vector $a_c = (x_a, y_a) = (3, 4)$ in Cartesian coordinates. Assume that these values are exact. What's the equivalent vector $a_p = (\theta_a, r_a)$ in the MPC plane? Express values to the nearest hundredth of a linear unit, and to the nearest degree.

SOLUTION

First, let's find the direction angle θ_a. Because $x_a > 0$ and $y_a > 0$, we can calculate

$$\theta_a = \text{Arctan}(y_a/x_a)$$
$$= \text{Arctan}(4/3)$$
$$= 53°$$

We can solve for r_a by doing the arithmetic

$$r_a = (x_a^2 + y_a^2)^{1/2}$$
$$= (3^2 + 4^2)^{1/2}$$
$$= (9 + 16)^{1/2}$$
$$= 25^{1/2}$$
$$= 5.00$$

When we put these resultant values together into an ordered pair, we get

$$a_p = (\theta_a, r_a)$$
$$= (53°, 5.00)$$

PROBLEM 7-6

Consider the vector $b_p = (\theta_b, r_b) = (200°, 4.55)$ in the MPC plane. Assume that these values are exact. Convert b_p to an equivalent vector $b_c = (x_b, y_b)$ in Cartesian coordinates. Express the answer to the nearest tenth of a unit for both coordinates x_b and y_b.

SOLUTION

Let's use the conversion formulas given a little while ago. First, with the aid of a scientific calculator, we solve for x_b to get

$$x_b = r_b \cos \theta_b$$
$$= 4.55 \cos 200°$$
$$= 4.55 \times (-0.9397)$$
$$= -4.3$$

Then we calculate y_b as

$$y_b = r_b \sin \theta_b$$
$$= 4.55 \sin 200°$$
$$= 4.55 \times (-0.3420)$$
$$= -1.6$$

Finally, we combine the resultants in an ordered pair to obtain

$$\mathbf{b}_c = (x_b, y_b)$$
$$= (-4.3, -1.6)$$

Vectors in Three Dimensions

In Cartesian xyz-space, we can denote or express vectors $\mathbf{a}$ and $\mathbf{b}$ as rays that run outward from the origin $(0,0,0)$ to points (x_a, y_a, z_a) and (x_b, y_b, z_b), as shown in Fig. 7-9.

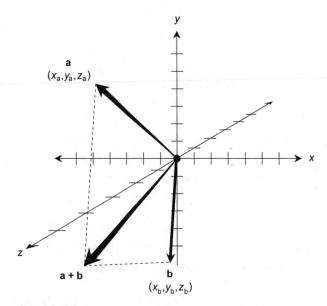

FIGURE 7-9 · Two vectors $\mathbf{a}$ and $\mathbf{b}$ in xyz-space. Their sum follows the diagonal of a parallelogram (distorted by perspective in this illustration).

Magnitude of a Vector

We can use a three-dimensional extension of the Pythagorean Theorem to calculate the magnitude of a vector **a**, written |**a**|, in Cartesian three-space. We square each coordinate value, add up the results, and then take the positive square root of that sum to get

$$|\mathbf{a}| = (x_a^2 + y_a^2 + z_a^2)^{1/2}$$

Direction of a Vector

In Cartesian three-space, we define the direction of a vector **a** in terms of the angles θ_x, θ_y, and θ_z that **a** subtends relative to the positive x, y, and z axes respectively (Fig. 7-10). These angles, expressed in radians as an ordered triple $(\theta_x, \theta_y, \theta_z)$, constitute the *direction angles* of **a**. Sometimes, mathematicians specify the cosines of these angles to get the so-called *direction cosines*. In that case, we have

$$\text{dir } \mathbf{a} = (\alpha, \beta, \gamma)$$

where $\alpha = \cos\theta_x$, $\beta = \cos\theta_y$, and $\gamma = \cos\theta_z$.

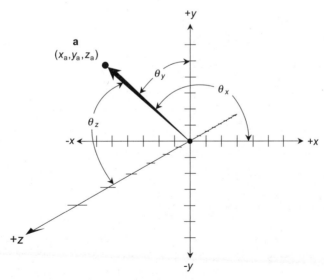

FIGURE 7-10 · Direction angles of a vector in *xyz*-space.

Sum of Two Vectors

Consider two vectors **a** and **b** in Cartesian three-space, defined in terms of the ordered triples

$$\mathbf{a} = (x_a, y_a, z_a)$$

and

$$\mathbf{b} = (x_b, y_b, z_b)$$

We can calculate the sum of the vectors by adding up their coordinate values individually, and then constructing an ordered triple out of the result to get

$$\mathbf{a} + \mathbf{b} = [(x_a + x_b), (y_a + y_b), (z_a + z_b)]$$

TIP   *As we do in the two-dimensional case, we can portray the sum of two vectors in Cartesian three-space by constructing a parallelogram with **a** and **b** as adjacent sides. The sum vector **a** + **b** runs along the diagonal of the parallelogram, as shown in Fig. 7-9. This diagram contains perspective, so the parallelogram appears distorted. Nevertheless, you should get the idea!*

Multiplying a Vector by Scalar

Imagine an arbitrary vector **a** in Cartesian three-space with the coordinates (x_a, y_a, z_a). If we multiply **a** by some positive real scalar k, we have

$$k\mathbf{a} = k(x_a, y_a, z_a)$$
$$= (kx_a, ky_a, kz_a)$$

If we multiply our vector **a** by some negative real scalar $-k$, then we get

$$-k\mathbf{a} = -k(x_a, y_a, z_a)$$
$$= (-kx_a, -ky_a, -kz_a)$$

Still Struggling

Imagine an arbitrary vector **a** whose direction angles we represent in terms of an ordered triple as

$$\text{dir } \mathbf{a} = (\theta_x, \theta_y, \theta_z)$$

We can express the direction angles of $k\mathbf{a}$ with the same ordered triple, so we have

$$\text{dir } k\mathbf{a} = (\theta_x, \theta_y, \theta_z)$$

The direction angles of vector $-k\mathbf{a}$, however, all increase or decrease by π, so we must represent them as

$$\text{dir } -k\mathbf{a} = [(\theta_x \pm \pi), (\theta_y \pm \pi), (\theta_z \pm \pi)]$$

Whether we should add π to an angle or subtract π from that angle depends on the value of the results we get in each case. Our result should always constitute a nonnegative angle measuring less than π (a half rotation in space, not a full rotation).

Dot Product of Two Vectors

The *dot product*, also known as the *scalar product* and written $\mathbf{a} \bullet \mathbf{b}$, of vectors $\mathbf{a}$ and $\mathbf{b}$ in Cartesian xyz-space is a real number given by the formula

$$\mathbf{a} \bullet \mathbf{b} = x_a x_b + y_a y_b + z_a z_b$$

where $\mathbf{a} = (x_a, y_a, z_a)$ and $\mathbf{b} = (x_b, y_b, z_b)$.

Alternatively, we can find the dot product $\mathbf{a} \bullet \mathbf{b}$ in terms of the vector magnitudes $|\mathbf{a}|$ and $|\mathbf{b}|$ along with the angle θ expressed counterclockwise going from $\mathbf{a}$ to $\mathbf{b}$ in the plane containing them both. That is,

$$\mathbf{a} \bullet \mathbf{b} = |\mathbf{a}| \, |\mathbf{b}| \cos \theta$$

Cross Product of Two Vectors

The *cross product*, also known as the *vector product* and written $\mathbf{a} \times \mathbf{b}$, of vectors $\mathbf{a}$ and $\mathbf{b}$ is a third vector that runs perpendicular to the plane containing $\mathbf{a}$ and $\mathbf{b}$. Suppose that θ represents the angle between vectors $\mathbf{a}$ and $\mathbf{b}$ going counterclockwise in the plane containing them both, as shown in Fig. 7-11. You can calculate the magnitude of the vector $\mathbf{a} \times \mathbf{b}$ with the formula

$$|\mathbf{a} \times \mathbf{b}| = |\mathbf{a}| \, |\mathbf{b}| \sin \theta$$

In this example, $\mathbf{a} \times \mathbf{b}$ points upward at a right angle to the plane containing, and defined by, the two vectors $\mathbf{a}$ and $\mathbf{b}$. If $0° < \theta < 180°$ $(0 < \theta < \pi)$, you can use the so-called *right-hand rule* to ascertain the direction of $\mathbf{a} \times \mathbf{b}$. Curl your

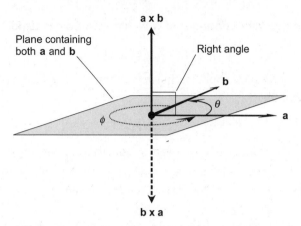

FIGURE 7-11 • The vector **b** × **a** has the same magnitude as vector **a** × **b**, but points in the opposite direction. Vectors **b** × **a** and **a** × **b** both run perpendicular to the plane containing (and defined by) the two original vectors **a** and **b**.

fingers in the sense in which θ, the angle between **a** and **b**, is defined. When you extend your thumb, the cross-product vector **a** × **b** points in the direction that your thumb goes.

When $180° < \theta < 360°$ ($\pi < \theta < 2\pi$), the cross-product vector reverses direction compared with the situation when $0° < \theta < 180°$ ($0 < \theta < \pi$). In the above formula for the vector magnitude, $\sin \theta$ is positive when $0° < \theta < 180°$ ($0 < \theta < \pi$), but negative when $180° < \theta < 360°$ ($\pi < \theta < 2\pi$). When $180° < \theta < 360°$ ($\pi < \theta < 2\pi$), you must use your left hand instead of your right hand, and curl your fingers into an almost complete circle! An example is the cross product **b** × **a** in Fig. 7-11. The angle ϕ, expressed counterclockwise between these vectors (as viewed from above), exceeds $180°$.

For any two vectors **a** and **b**, the vector **b** × **a** forms a "mirror image" of **a** × **b**, where the "mirror" corresponds to the plane containing both vectors. You can envision the "mirror image" if you imagine that **b** × **a** has the same magnitude as **a** × **b**, but points in exactly the opposite direction. Alternatively, you might say that the direction of **b** × **a** is the same as the direction of **a** × **b**, but the magnitudes of the two vectors are negatives of each other. The cross-product operation is *anticommutative*, meaning that

$$a \times b = -(b \times a)$$

Still Struggling

The absolute value of a scalar (a "plain old real number") can never be negative, but with vectors, negative magnitudes occasionally appear in equations. Whenever you encounter a vector whose magnitude turns out negative in your calculations, you should think of that vector as the equivalent of a positive vector pointing in the opposite direction. For example, if you're working out a physics or engineering problem and your calculations tell you that a certain machine produces a force of −20 newtons going straight up, you can think of that force vector as 20 newtons going straight down.

 PROBLEM 7-7

What's the magnitude of the vector denoted by $\mathbf{a} = (x_a, y_a, z_a) = (1, 2, 3)$? Consider the values 1, 2, and 3 as mathematically exact quantities. Express the answer to four decimal places.

 SOLUTION

You can use the formula for the length a vector in Cartesian xyz-space to calculate the magnitude as

$$|\mathbf{a}| = (x_a^2 + y_a^2 + z_a^2)^{1/2}$$
$$= (1^2 + 2^2 + 3^2)^{1/2}$$
$$= (1 + 4 + 9)^{1/2}$$
$$= 14^{1/2}$$
$$= 3.7417$$

 PROBLEM 7-8

Consider two vectors $\mathbf{a}$ and $\mathbf{b}$ in Cartesian xyz-space, both of which lie in the xy-plane. Suppose that the vectors have the ordered triples

$$\mathbf{a} = (3, 4, 0)$$

and

$$\mathbf{b} = (0, -5, 0)$$

Assume that these coordinate values are mathematically exact. Find the ordered triple representing the vector $a \times b$.

SOLUTION

Envision the two vectors a and b as they appear in the xy-plane. Figure 7-12 shows the situation. In this drawing, imagine the positive z axis coming out of the page directly toward you, and the negative z axis pointing straight away from you on the other side of the page.

The cross product of two vectors always runs perpendicular to the plane containing the original vectors. In this case, $a \times b$ points along the z axis. The ordered triple must therefore have the form $(0, 0, z)$, where z represents a real number. You don't yet know the exact value of z, but you'll find out pretty soon!

Calculate the lengths (magnitudes) of the two vectors a and b. To determine $|a|$, use the formula

$$|a| = (x_a^2 + y_a^2)^{1/2}$$
$$= (3^2 + 4^2)^{1/2}$$
$$= (9 + 16)^{1/2}$$
$$= 25^{1/2}$$
$$= 5$$

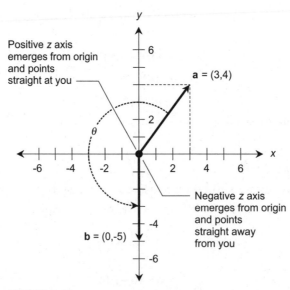

FIGURE 7-12 · Illustration for Problem 7-8.

To calculate $|\mathbf{b}|$, use the formula

$$|\mathbf{b}| = (x_b^2 + y_b^2)^{1/2}$$
$$= [0^2 + (-5)^2]^{1/2}$$
$$= 25^{1/2}$$
$$= 5$$

Now that you know $|\mathbf{a}| = 5$ and $|\mathbf{b}| = 5$, you can calculate

$$|\mathbf{a}||\mathbf{b}| = 5 \times 5$$
$$= 25$$

In order to determine the magnitude of $\mathbf{a} \times \mathbf{b}$, you must multiply 25 by the sine of the angle θ going counterclockwise from $\mathbf{a}$ to $\mathbf{b}$. Notice that the measure of θ equals 270° (three-quarters of a circle) minus the angle between the x axis and vector $\mathbf{a}$. The angle between the x axis and vector $\mathbf{a}$ equals the arctangent of the y-coordinate of $\mathbf{a}$ divided by the x-coordinate of $\mathbf{a}$. That's arctan 4/3, or approximately 53° as determined using a calculator. Therefore

$$\theta = 270° - 53°$$
$$= 217°$$

Now that you know the angle, you can use a calculator to determine that

$$\sin \theta = \sin 217°$$
$$= -0.60$$

Recall the formula for the magnitude of a cross-product vector as

$$|\mathbf{a} \times \mathbf{b}| = |\mathbf{a}||\mathbf{b}| \sin \theta$$

In this case,

$$|\mathbf{a} \times \mathbf{b}| = 25 \times (-0.60)$$
$$= -15$$

In this result, the minus sign tells you that the cross product vector points negatively along the z axis, or straight away from you, as it would appear in Fig. 7-12. The z coordinate of $\mathbf{a} \times \mathbf{b}$ equals −15. You know that the x and y coordinates of $\mathbf{a} \times \mathbf{b}$ both equal 0 because $\mathbf{a} \times \mathbf{b}$ lies precisely along the z axis. You can put all this knowledge together to conclude that

$$\mathbf{a} \times \mathbf{b} = (0,0,-15)$$

TIP   *The foregoing calculations are accurate to two significant figures. However, if you use a good scientific calculator (such as the one provided with most personal computers) and let it go to all the significant figures it can handle, you'll discover that the answer is mathematically exact. You can* precisely *state that*

$$\mathbf{a} \times \mathbf{b} = (0,0,-15)$$

provided that the original two vectors precisely *equal*

$$\mathbf{a} = (3,4,0)$$

and

$$\mathbf{b} = (0,-5,0)$$

QUIZ

Refer to the text in this chapter if necessary. A good score is eight correct. Answers are in the back of the book.

1. **One minute (1 m) of celestial right ascension equals**
    A. 1/60 of a full circle.
    B. 1/360 of a full circle.
    C. 1/720 of a full circle.
    D. 1/1440 of a full circle.

2. **Figure 7-13 illustrates a point P plotted in a system of**
    A. celestial coordinates.
    B. cylindrical coordinates.
    C. spherical coordinates.
    D. terrestrial coordinates.

3. **In Fig. 7-13, the point P′ lies in the xy-plane, directly underneath point P so that the line segment PP′ runs parallel to the z axis. Each radial division (the distance between each concentric coordinate circle in the xy-plane) equals 5 units. Point P′ lies 18 units from the origin, and point P is 24 units above point P′. Based on this information, how far from the origin does P lie, as measured along a straight line?**
    A. 27 units
    B. 30 units
    C. 36 units
    D. We need more information to say.

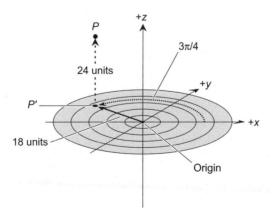

FIGURE 7-13 · Illustration for Quiz Questions 2 through 5.

4. Based on the information given in Fig. 7-13 and the statement of Question 3, what are the coordinates of P in Cartesian xyz-space, rounded off to two decimal places? Here's a hint: Find the coordinates of P' first; think of the xy-plane as coincident with a two-dimensional polar coordinate plane, about which you learned in Chap. 6. Here's another hint: Consider the distances and angle in this figure as mathematically exact values.

   A. (18.00 , 18.00 , 24.00)
   B. (13.42 , 13.42 , 24.00)
   C. (−12.73, 12.73, 24.00)
   D. (12.00 , −12.00 , −24.00)

5. Imagine a line segment running straight out from the origin to the point P in the scenario of Fig. 7-13. What angle, expressed to the nearest degree, does this line segment subtend in three-space relative to the line segment connecting the origin to point P'? Here's a hint: You'll need to use a little of the knowledge of inverse circular functions that you gained in Chap. 4.

   A. 45°
   B. 60°
   C. 48°
   D. 53°

6. What's the magnitude of the vector $\mathbf{a} = (5,-7)$ in the Cartesian xy-plane, accurate to three decimal places? Assume that the coordinate values are exact.

   A. 8.602
   B. 6.000
   C. 5.916
   D. We need more information to say.

7. What's the MPC direction angle of the vector $\mathbf{a} = (5,-7)$ in the Cartesian plane, accurate to the nearest degree? Assume that the coordinate values are exact.

   A. 306°
   B. 234°
   C. 216°
   D. 54°

8. What are the Cartesian xy-coordinates of the MPC vector $\mathbf{a} = (7\pi/6, 5.00)$? Assume that these MPC values are exact. Express the Cartesian values to the nearest hundredth of a unit. Consider the value of π as 3.14159.

   A. (4.33 , 2.50)
   B. (−4.33 , 2.50)
   C. (−4.33 , −2.50)
   D. (4.33 , −2.50)

9. Suppose that you want to express the MPC vector $\mathbf{a} = (7\pi/6, 5.00)$ in NPC terms instead. Assume that these MPC values are exact. Express the angle to the nearest degree, and the radius to the nearest hundredth of a unit. Consider the value of π as 3.14159. Here's a hint: Look back at the text in Chap. 6 if you get confused.

A. $(150°, 5.00)$

B. $(210°, 5.00)$

C. $(240°, 5.00)$

D. $(300°, 5.00)$

10. What's the magnitude of the vector denoted by $\mathbf{a} = (x,y,z) = (5,-3,-2)$? Consider all three coordinate values as mathematically exact quantities. Express the answer to the nearest hundredth of a unit.

A. 7.00

B. 6.74

C. 6.36

D. 6.16

Test: Part I

Do not refer to the text when taking this test. You may draw diagrams or use a calculator if necessary. A good score is at least 38 correct. Answers are in the back of the book. It's best to have a friend check your score the first time, so you won't memorize the answers if you want to take the test again.

1. Which of the following terms refers to a mapping between two sets in which each and every element of one set corresponds to exactly one element of the other set (working both ways)?
   A. Injection
   B. Bijection
   C. Surjection
   D. Rejection
   E. Projection

2. The hyperbolic functions operate on
   A. angles between the positive x axis and points on a hyperbola.
   B. distances between points on a hyperbola.
   C. areas of enclosed regions adjacent to a hyperbola.
   D. angles between asymptotes of a hyperbola.
   E. perimeters of triangles with vertices on a hyperbola.

3. Figure Test I-1 illustrates a right triangle where θ represents the measure of the interior angle between the base and the hypotenuse. The ratio of the base length to the hypotenuse length equals
   A. sin θ.
   B. cos θ.
   C. tan θ.
   D. csc θ.
   E. None of the above

4. The ratio of the height to the base length in Fig. Test I-1 equals
   A. sin θ.
   B. cos θ.
   C. tan θ.
   D. csc θ.
   E. None of the above

5. The ratio of the hypotenuse length to the height in Fig. Test I-1 equals
   A. sin θ.
   B. cos θ.

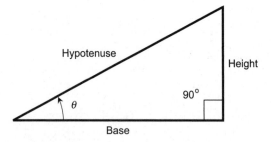

FIGURE TEST I-1 · Illustration for Part I Test Questions 3 through 6.

   C. tan θ.
   D. csc θ.
   E. None of the above

6. **The ratio of the base length to the height in Fig. Test I-1 equals**
   A. sin θ.
   B. cos θ.
   C. tan θ.
   D. csc θ.
   E. None of the above

7. **The domain of the cosine function extends over the set of**
   A. all real numbers.
   B. positive real numbers only.
   C. all real numbers between and including −1 and 1.
   D. all real numbers between and including −π/2 and π/2.
   E. All real numbers except 0.

8. **The expression $z \in (-3,3]$ tells us that**
   A. $-3 < z < 3$.
   B. $-3 < z \leq 3$.
   C. $-3 \leq z < 3$.
   D. $-3 \leq z \leq 3$.
   E. $z < -3$ or $3 < z$.

9. **The expression $z \in [-3,3]$ tells us that**
   A. $-3 < z < 3$.
   B. $-3 < z \leq 3$.
   C. $-3 \leq z < 3$.
   D. $-3 \leq z \leq 3$.
   E. $z \neq -3$ and $z \neq 3$.

10. **Mathematicians and engineers sometimes calculate values of the hyperbolic sine in terms of**
   A. exponential functions.
   B. quadratic functions.
   C. circular functions.
   D. linear functions.
   E. polar coordinates.

11. **According to the unit-circle paradigm, two different angles both have tangents equal to exactly 1. What are those angles?**
   A. 90° and 270°
   B. 45° and 225°
   C. 60° and 240°
   D. 30° and 210°
   E. 135° and 315°

12. **How many radians does a half-circle arc contain?**

    A. $\pi/4$

    B. $\pi/2$

    C. $3\pi/4$

    D. π

    E. 2π

13. **What's the radian equivalent of a 135° angle?**

    A. $3\pi/8$

    B. $5\pi/6$

    C. $3\pi/4$

    D. $2\pi/3$

    E. $3\pi/2$

14. **The function $y = \tan x$ passes through an inflection point when**

    A. $x = 0°$.

    B. $x = 45°$.

    C. $x = 90°$.

    D. $x = 135°$.

    E. All of the above

15. **The hyperbolic cosine is equivalent to the**

    A. reciprocal of the hyperbolic secant.

    B. ratio of the hyperbolic sine to the hyperbolic secant.

    C. reciprocal of the hyperbolic tangent.

    D. ratio of the hyperbolic tangent to the hyperbolic cotangent.

    E. reciprocal of the hyperbolic sine.

16. **In Fig. Test I-2, how far does point P lie from the origin?**

    A. The square root of 12 units

    B. 7 units

    C. 5 units

    D. 9/2 units

    E. We need more information to say.

17. **In Fig. Test I-2, how far does point Q lie from the origin?**

    A. 10 units

    B. The square root of 50 units

    C. 15/5 units

    D. 13/5 units

    E. We need more information to say.

18. **In Fig. Test I-2, how far does point P lie from point Q?**

    A. The square root of 145 units

    B. The square root of 120 units

FIGURE TEST I-2 · Illustration for Part I Test Questions 16 through 19.

C. 5 plus the square root of 50 units
D. 19/2 units
E. We need more information to say.

19. **In Fig. Test I-2, what are the coordinates of the midpoint of the line segment connecting points P and Q?**
    A. $(-1/2,-1)$
    B. $(-1,-1/2)$
    C. $(0,-1/2)$
    D. $(-3/2,-3/2)$
    E. We need more information to say.

20. **In Cartesian uv-coordinates where u represents the independent variable and v represents the dependent variable, we can define the hyperbolic functions on the basis of a curve having one of the following equations. Which one?**
    A. $u^2 + v^2 = 1$
    B. $v = u^2$
    C. $v = 1/(2u)$
    D. $u^2 - v^2 = 1$
    E. $u^2 + v^2 = -1$

21. **The range of the tangent function extends over the set of**
    A. all real numbers.
    B. positive real numbers only.
    C. all real numbers between and including −1 and 1.
    D. all real numbers between and including $-\pi/2$ and $\pi/2$.
    E. All real numbers except 0.

22. **In a polar coordinate plane, we define the position of a point on the basis of**
    A. two distances.
    B. a distance and an angle.
    C. two angles.
    D. two angles and a distance.
    E. two distances and an angle.

23. **What are the Cartesian *xy*-plane coordinates of the mathematician's polar coordinate (MPC) point $(\theta, r) = (\pi, 2^{1/2})$?**
    A. (1,1)
    B. (−1,1)
    C. (−1,−1)
    D. (1,−1)
    E. None of the above

24. **At which of the following values of *x* can we *not define* the function $y = \sin x$ in the unit-circle paradigm?**
    A. $x = 0$
    B. $x = \pi/4$
    C. $x = \pi/3$
    D. $x = 7\pi/4$
    E. None of the above. We can define $y = \sin x$ for all the foregoing values of *x* in the unit-circle paradigm.

25. **Figure Test I-3 is a graph of the function $y = \text{Arcsec } x$. The left-hand curve extends forever to the left, converging on (but never reaching) the horizontal line $y = \pi/2$ from the positive side. The right-hand curve extends forever to the right, converging on (but never reaching) the line $y = \pi/2$ from the negative side. Based on this description and the general appearance of the graph, the domain of this function contains all real numbers**

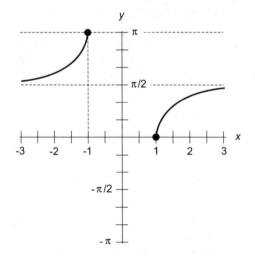

FIGURE TEST I-3 · Illustration for Part I Test
Questions 25 and 26.

A. except those in the open interval (−1,1).
B. in the open interval (−1,1).
C. except π/2.
D. in the half-open interval [0, π/2) or the half-open interval (π/2, π].
E. without exception.

26. **The range of the function shown in Fig. Test I-3 contains all real numbers**
    A. except those in the open interval (−1,1).
    B. in the open interval (−1,1).
    C. except π/2.
    D. in the half-open interval [0, π/2) or the half-open interval (π/2, π].
    E. without exception.

27. **Mathematicians and engineers sometimes calculate values of the hyperbolic Arccosine in terms of**
    A. circular functions.
    B. quadratic functions.
    C. logarithmic functions.
    D. inverse quadratic functions.
    E. polar coordinates.

28. **Suppose that we see an equation where a dependent variable *r* is a function of an independent variable *θ* such that**

$$r = -\theta/\pi$$

    If we plot a graph of this equation in mathematician's polar coordinates (MPC), we'll get a
    A. pair of spirals.
    B. circle centered at the origin.
    C. circle passing through the origin.
    D. straight line passing through the origin.
    E. straight line that doesn't pass through the origin.

29. **Figure Test I-4 is a graph of the hyperbolic cotangent function *y* = coth *x*. The domain contains**
    A. all real numbers.
    B. all positive real numbers.
    C. all negative real numbers.
    D. all real numbers except 0.
    E. all real numbers except those in the interval [−1,1].

30. **Based on the graph of Fig. Test I-4, as *x* grows larger *negatively* without limit, the value of coth *x***
    A. approaches −1 from below (the negative side).
    B. approaches −1 from above (the positive side).
    C. approaches 0 from the left (the negative side).
    D. approaches 0 from the right (the positive side).
    E. increases positively without limit.

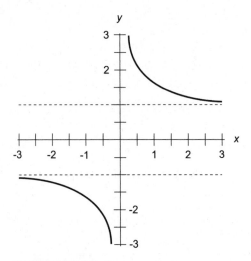

FIGURE TEST I-4 · Illustration for Part I Test Questions 29 and 30.

31. In the unit-circle paradigm, two different angles can both have cotangents equal to exactly −1. What are those angles?
   A. $\pi/2$ and $3\pi/2$
   B. $\pi/4$ and $5\pi/4$
   C. $\pi/3$ and $4\pi/3$
   D. $\pi/6$ and $5\pi/6$
   E. $3\pi/4$ and $7\pi/4$

32. A celestial latitude coordinate can equal any value within the interval
   A. $[-180°,+180°]$.
   B. $[-90°,+90°]$.
   C. $(0°,+90°)$.
   D. $[0°,+180°)$.
   E. $[0°,+360°)$.

33. Consider two vectors that run perpendicular to each other. One of them measures 9 units long, and the other one measures 4 units long. What's their dot product?
   A. 36
   B. 6
   C. 0
   D. $36 \times 2^{1/2}$
   E. We need more information to answer this question.

34. What's the magnitude of the vector $(-1,0,1)$ in Cartesian xyz-space?
   A. $2^{1/2}$
   B. $3^{1/2}$

C. 2

D. 3

E. We need more information to answer this question.

35. **We can always define a mapping between the elements of two different number sets in terms of**

   A. inverses.

   B. surjections.

   C. functions.

   D. ordered pairs.

   E. codomains.

36. **Someone tells us that a certain angle inside a triangle measures 15 minutes of arc. That's the equivalent of**

   A. 1/240 of an angular degree.

   B. 1/180 of an angular degree.

   C. 1/15 of an angular degree.

   D. 1/4 of an angular degree.

   E. 4 angular degrees.

37. **Figure Test I-5 shows a straight line L and a circle C in the mathematician's polar coordinate (MPC) plane. What's the equation of line L? Assume that each radial increment (distance between any pair of concentric circles) represents 1 unit.**

   A. $r = 3\pi/4$

   B. $\theta = 3\pi/4$

   C. $\theta = 4$

   D. $r = \pi/6$

   E. $\theta = \pi/4$

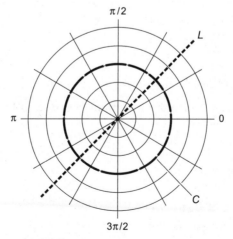

FIGURE TEST I-5 · Illustration for Part I Test Questions 37 and 38.

38. What's the MPC equation of circle C in Fig. Test I-5? Assume that each radial increment (distance between any pair of concentric circles) represents 1 unit.

   A. $r = 3\pi$
   B. $r = \pi/3$
   C. $\theta = 3$
   D. $r = 3$
   E. $\theta = 3\pi/4$

39. Consider the point $(x,y) = (7,0)$ in a Cartesian coordinate plane where x represents the independent variable and y represents the dependent variable. What are the coordinates (θ,r) of this point in the mathematician's polar coordinate (MPC) plane where θ represents the direction and r represents the radius?

   A. $(\pi/4 , 98^{1/2})$
   B. $(7,0)$
   C. $(0,7)$
   D. $(7\pi,0)$
   E. $[(49\pi)^{1/2} , -(49\pi)^{1/2}]$

40. Consider the point $(x,y) = (7,7)$ in a Cartesian coordinate plane where x represents the independent variable and y represents the dependent variable. What are the coordinates (θ,r) of this point in the mathematician's polar coordinate (MPC) plane where θ represents the direction and r represents the radius?

   A. $(\pi/4 , 98^{1/2})$
   B. $(7,0)$
   C. $(0,7)$
   D. $(7\pi,0)$
   E. $[(49\pi)^{1/2} , -(49\pi)^{1/2}]$

41. Suppose that we draw a graph of the circle $x^2 + y^2 = 1$ on the Cartesian xy-plane. Then we choose a point on the circle and draw a ray outward from the origin passing through that point, such that the ray subtends a certain angle relative to the positive x axis. The point's abscissa (x-coordinate) equals the

   A. sine of the angle.
   B. cosine of the angle.
   C. secant of the angle.
   D. cosecant of the angle.
   E. tangent of the angle.

42. In the situation described in the statement of Question 41, the point's ordinate (y-coordinate) equals the

   A. sine of the angle.
   B. cosine of the angle.
   C. secant of the angle.
   D. cosecant of the angle.
   E. tangent of the angle.

43. In Fig. Test I-6, vector c could illustrate

   A. $\mathbf{a} \cdot \mathbf{b}$.
   B. $\mathbf{b} \cdot \mathbf{a}$.

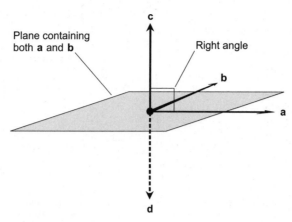

FIGURE TEST I-6 • Illustration for Part I Test Questions 43 and 44.

   C. **a** × **b**.
   D. **b** × **a**.
   E. nothing that we've learned about in this course.

44. **In Fig. Test I-6, vector d (which runs in the opposite direction from vector c) could illustrate**
   A. **a** • **b**.
   B. **b** • **a**.
   C. **a** × **b**.
   D. **b** × **a**.
   E. nothing that we've learned about in this course.

45. **What's the degree equivalent of an angle whose radian measure equals $7\pi/4$?**
   A. 240°
   B. 270°
   C. 285°
   D. 300°
   E. 315°

46. **Suppose that we see an equation where a dependent variable r is a function of an independent variable θ such that**

$$4r = 5\pi$$

If we plot a graph of this equation in mathematician's polar coordinates (MPC), we'll get a
   A. pair of spirals.
   B. circle centered at the origin.
   C. circle passing through the origin.
   D. straight line passing through the origin.
   E. straight line that doesn't pass through the origin.

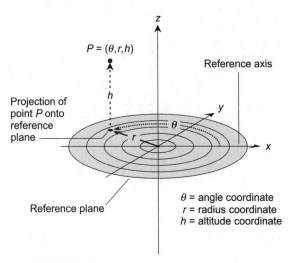

FIGURE TEST I-7 · Illustration for Part I Test Question 48.

47. Imagine two points $R = (x_r, y_r)$ and $S = (x_s, y_s)$ in the Cartesian coordinate plane. What are the coordinates of the midpoint M of a line segment connecting R and S?

A. $M = [(x_r + y_r), (x_s + y_s)]$
B. $M = (x_r y_r, x_s y_s)$
C. $M = [(x_r + y_s), (x_s + y_r)]$
D. $M = (x_r y_s, x_s y_r)$
E. $M = \{[(x_r + x_s)/2], [(y_r + y_s)/2]\}$

48. Figure Test I-7 illustrates the basic scheme for a system of

A. spherical coordinates.
B. cylindrical coordinates.
C. latitude and longitude coordinates.
D. Cartesian three-space coordinates.
E. elliptical coordinates.

49. Imagine that someone defines the location of a point in space by specifying the point's azimuth, elevation, and range. This person refers to a system of

A. Cartesian coordinates.
B. cylindrical coordinates.
C. spherical coordinates.
D. navigator's polar coordinates.
E. right ascension and declination coordinates.

50. Fill in the blank to make the following statement true: "We define a function as a relation in which every element in the domain corresponds to _____ in the range."

A. at least one element
B. at most one element
C. exactly one element
D. two or more elements
E. infinitely many elements

Part II

Extensions and Applications

chapter **8**

Scientific Notation

In some applications, trigonometry involves vast distances and tiny angles that don't lend themselves to expression as ordinary numerals. In such cases, we can use *scientific notation*, also called *power-of-10 notation*, to keep things from getting cumbersome.

CHAPTER OBJECTIVES

In this chapter, you will

- Learn to properly write and interpret subscripts and superscripts.
- Denote large and small numbers using powers of 10.
- Define orders of magnitude and learn the basic prefix multipliers.
- Decide when you should use scientific notation (and when you shouldn't).
- Estimate and express precision and error.
- Determine the accuracy of a calculation based on significant figures.

Formatting Methods

Subscripts modify the meanings of units, constants, and variables. We place a subscript immediately to the right of a character and slightly below it. *Superscripts* usually represent *exponents* (raising to a power). We place a superscript immediately to the right of a character and slightly above it.

Examples of Subscripts

You should never italicize numeric subscripts. Alphabetic subscripts can appear in italics if they represent variables. Here are some expressions with subscripts:

θ_0	read "theta sub nought"; stands for a specific angle
R_{out}	read "R sub out"; stands for output resistance in an electronic circuit
y_n	read "y sub n"; represents a variable with a variable subscript

You'll rarely, if ever, have any reason to modify a numeral with a subscript. Don't expect to see or write expressions such as 3_5, -9.7755_π, or 16_x.

Some physical constants include subscripts. An example is m_e, representing the mass of an electron at rest. (The "e" in this context does not appear italicized because it stands for the word "electron," not for a mathematical quantity.)

Still Struggling

Subscripts offer notational clarity when things would get confusing without them. For example, you can denote points in space by writing ordered triples such as (x_1, x_2, x_3) rather than (x, y, z). This subscripting scheme becomes especially useful if you want to write about points in a higher-dimensional space, for example, $(x_1, x_2, x_3, \dots, x_{11})$ in Cartesian 11-dimensional (11D) space.

Examples of Superscripts

You should never italicize numeric superscripts, but you will want to italicize alphabetic superscripts in most cases. Here are some expressions with superscripts:

2^3 read "two cubed";
         represents $2 \times 2 \times 2$

$\sin^2 \theta$ read "the square of the sine of theta";
         represents a quantity multiplied by itself

$\text{Sin}^{-1} \theta$ read "the inverse sine of theta";
         alternative expression for Arcsin θ

Standard Power-of-10 Notation

We denote a specific numeral in *standard power-of-10 notation* as the product in the form

$$m.n \times 10^z$$

where the dot (.) is a period written on the base line (not a raised dot indicating multiplication). It represents the *radix point* or *decimal point*. The value m (to the left of the radix point) is a positive or negative single digit. The value n (to the right of the radix point) is a nonnegative integer. The entire quantity to the left of the multiplication symbol constitutes the *coefficient*. The value z, which tells us the power of 10, is an integer. Following are some examples of numbers written in standard scientific notation:

$$2.56 \times 10^6$$

$$8.0773 \times 10^{-18}$$

$$1.000 \times 10^0$$

$$-5.45 \times 10^{12}$$

Alternative Power-of-10 Notation

Once in awhile, you'll encounter *alternative power-of-10 notation*, where no nonzero numerals ever appear to the left of the decimal point. We might see a single cipher there, or we might see nothing at all (the whole expression will begin with the decimal point). When we express the above quantities in the alternative power-of-10 form, they're decimal fractions larger than 0 but less than 1, and the value of the exponent increases by 1 compared with the standard form, so we would write

$$0.256 \times 10^7$$

$$0.80773 \times 10^{-17}$$

$$0.1000 \times 10^1$$

$$-0.545 \times 10^{13}$$

or perhaps

$$.256 \times 10^7$$

$$.80773 \times 10^{-17}$$

$$.1000 \times 10^1$$

$$-.545 \times 10^{13}$$

TIP   *The above four quantities have values identical to their counterparts in the preceding paragraph. They're expressed differently, that's all. It's like saying that you should cook a hamburger for a quarter of an hour instead of for 15 minutes.*

The "Times Sign"

Most scientists in America use the cross symbol ($\times$), as in the foregoing examples, to denote the multiplication inside a power-of-10 quantity. But a small dot raised above the base line ($\cdot$) can also play this role, so the above numbers look like this:

$$2.56 \cdot 10^6$$

$$8.0773 \cdot 10^{-18}$$

$$1.000 \cdot 10^0$$

$$-5.45 \cdot 10^{12}$$

Don't confuse this small dot with a radix point! In the general expression

$$m.n \cdot 10^z$$

the dot between m and n represents a radix point and lies along the base line, while the dot between n and 10^z represents multiplication and lies above the base line.

When using an old-fashioned typewriter, or in word processors that lack a good repertoire of symbols, the lowercase, non-italicized letter x can indicate multiplication, as long as readers know that you don't intend for it to stand as a variable (as it would if it were italicized). You can also use an asterisk (*) to denote multiplication, as follows:

$$2.56 * 10^6$$

$$8.0773 * 10^{-18}$$

$$1.000 * 10^0$$

$$-5.45 * 10^{12}$$

Plain-Text Exponents

Once in awhile, you'll have to express numbers in power-of-10 notation using plain, unformatted text. This sort of situation might arise when you transmit information within the body of an e-mail message (as opposed to sending a formatted file as an attachment). Some calculators employ plain-text notation. An uppercase or lowercase letter E indicates that the quantity immediately following is a power of 10. The power-of-10 designator always includes a sign (plus or minus) unless it happens to equal 0. In this format, you would write the above quantities as

$$2.56E{+}6$$
$$8.0773E{-}18$$
$$1.000E0$$
$$-5.45E{+}12$$

or as

$$2.56e{+}6$$
$$8.0773e{-}18$$
$$1.000e0$$
$$-5.45e{+}12$$

Some people, especially computer programmers, use an asterisk (*) to indicate multiplication and an inverted wedge (^) to indicate a superscript, so the preceding expressions become

$$2.56 * 10 \wedge 6$$
$$8.0773 * 10 \wedge {-}18$$
$$1.000 * 10 \wedge 0$$
$$-5.45 * 10 \wedge 12$$

Still Struggling

Regardless of which notational variant you prefer for scientific notation, the above-quoted numerical values, written out in expanded decimal form, are

$$2{,}560{,}000$$
$$0.000000000000000080773$$
$$1.000$$
$$-5{,}450{,}000{,}000{,}000$$

Orders of Magnitude

As you can see, power-of-10 notation allows us to easily write down numbers that denote huge or tiny values. In fact, we can use the notation to quantify things that would dwarf the universe, or that would make a subatomic particle look huge by comparison. For example, consider

$$2.55 \times 10^{45,589}$$

and

$$-9.8988 \times 10^{-7,654,321}$$

Imagine the task of writing either of these numbers out as sequences of digits in ordinary decimal form! In the first case, you'd have to write the numerals 255, and then follow them with a string of 45,587 ciphers (numerals 0). In the second case, you'd have to write a minus sign, then a numeral 0, then a radix point, then a string of 7,654,320 ciphers, then the numerals 9, 8, 9, 8, and 8. Now consider

$$2.55 \times 10^{45,592}$$

and

$$-9.8988 \times 10^{-7,654,318}$$

Both of these new numbers are *1000 times as large* as the previous two. You can tell by looking at the exponents. Both exponents have increased by 3. The number 45,592 equals 45,589 + 3, and the number −7,754,318 equals −7,754,321 + 3.

TIP   *In the strict mathematical sense, numbers grow larger as they get more positive or less negative; numbers grow smaller as they get more negative or less positive.*

In the scientific-notation examples given above, the second two numbers both exceed the first two numbers by three *orders of magnitude* (powers of 10). The written symbols look almost the same in either case, but they're as different as a meter compared to a kilometer, or a gram compared to a kilogram, or a year compared to a millennium.

The order-of-magnitude concept allows us to construct number lines, charts, and graphs with scales that cover huge spans of values. Figure 8-1 illustrates three examples:

1. Drawing A shows a number line spanning three orders of magnitude, from 1 to 1000.

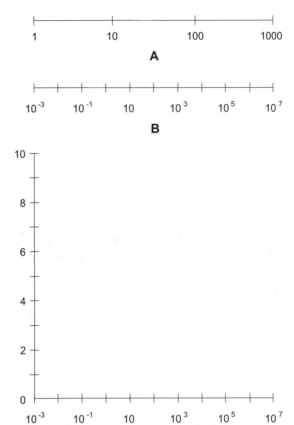

FIGURE 8-1 · At A, a number line spanning three orders of magnitude. At B, a number line spanning 10 orders of magnitude. At C, a two-dimensional coordinate system whose horizontal scale spans 10 orders of magnitude, and whose vertical scale extends from 0 to 10.

2. Drawing B shows a number line spanning 10 orders of magnitude, from 10^{-3} to 10^7.

3. Drawing C portrays a two-dimensional graph whose horizontal scale spans 10 orders of magnitude, from 10^{-3} to 10^7, and whose vertical scale extends in a linear manner from 0 to 10.

TIP *If you're astute, you'll notice that while the 0-to-10 linear scale "looks simple" and "reads easy," it covers infinitely many orders of magnitude. In that sense, it's esoteric indeed! This counterintuitive situation arises from the fact that, no matter how many times you divide a nonzero number by 10 (reduce it by an order of magnitude), you'll never get all the way down to 0.*

Prefix Multipliers

Physicists and engineers use verbal *prefix multipliers* to express orders of magnitude in power-of-10 notation. Table 8-1 lists the standard prefix multipliers for factors ranging from 10^{-24} to 10^{24}. Do you recognize any of these "beasts" from your past mathematical adventures? You should have little trouble thinking of some situations where you've read or heard these multipliers. For example:

- Visible light has a wavelength on the order of 400 to 700 *nano*meters (multiples of 10^{-9} meter).

TABLE 8-1 Power-of-10 prefix multipliers and their abbreviations.

Designator	Symbol	Multiplier
yocto–	y	10^{-24}
zepto–	z	10^{-21}
atto–	a	10^{-18}
femto–	f	10^{-15}
pico–	p	10^{-12}
nano–	n	10^{-9}
micro–	μ or mm	10^{-6}
milli–	m	10^{-3}
centi–	c	10^{-2}
deci–	d	10^{-1}
(none)	—	10^{0}
deka–	da or D	10^{1}
hecto–	h	10^{2}
kilo–	K or k	10^{3}
mega–	M	10^{6}
giga–	G	10^{9}
tera–	T	10^{12}
peta–	P	10^{15}
exa–	E	10^{18}
zetta–	Z	10^{21}
yotta–	Y	10^{24}

- You might take a pill with 400 *micro*grams (multiples of 10^{-6} gram) of folic acid.

- An English inch equals approximately 25.4 *milli*meters (multiples of 10^{-3} meter).

- The moon orbits at a distance of about 4×10^5 *kilo*meters (multiples of 10^3 meters) from the earth.

- Your Internet connection operates at a certain number of *mega*bits per second (multiples of 10^6 bits per second).

- Your computer has a processor with a frequency of a certain number of *giga*hertz (multiples of 10^9 cycles per second).

PROBLEM 8-1

Express the angle 5″ in scientific notation. Then determine the difference, in orders of magnitude, between the angles 180° and 5″. Assume that both angle values are exact. Calculate the answers to an accuracy of three decimal places.

✔SOLUTION

We recall that the accent symbol (′) means one minute of arc, or 1/60 of an angular degree, and the double accent (″) means one second of arc, which equals 1/60 of an arc minute or 1/3600 of an angular degree. Therefore

$$5'' = 5/3600$$

$$= 0.00138888\ldots$$

$$= 1.38888\ldots \times 10^{-3}$$

$$= (1.389 \times 10^{-3})°$$

The ellipsis (…) indicates that more digits follow the ones shown, but we can disregard them as "needlessly precise." Now let's divide 180° by (5/3600)° to determine the size of the large angle compared to the tiny one. We calculate

$$180°/(5/3600)° = 180 \times (3600/5)$$

$$= 180 \times 720$$

$$= 129{,}600$$

$$= 1.296 \times 10^5$$

An angle of 180° exceeds an angle of 5″ by roughly five orders of magnitude.

PROBLEM 8-2

What's the sine of 5″? Express the result to an accuracy of three decimal places.

SOLUTION

This problem requires a calculator that can work in scientific notation, defaulting to powers of 10 for extreme values. Most personal computers have calculators of this type. Before we do any work, we must set the calculator to work in degrees, not radians. Then we convert the angle to a decimal number of degrees, as we did in the first part of Problem 8-1, leaving in the entire string of digits (not rounding off). We have

$$5″ = 5/3600$$
$$= 0.00138888\ldots°$$

Now we can hit the "sin" button and round the result off. My calculator comes up with

$$\sin 0.00138888\ldots° = 2.424068405310\ldots e{-}5$$

which we can round off and write down as

$$= 2.424 \times 10^{-5}$$

TIP   *In science and engineering, you should use power-of-10 notation to denote a quantity when the power of 10 exceeds 3 or is smaller than −3. If the power of 10 equals any integer between −3 and 3 inclusive, you can (and normally should) denote a quantity as a "plain old number."*

Rules for Use

Let's see how power-of-10 notation works in arithmetic involving addition, subtraction, multiplication, division, exponentiation, and roots.

Addition

When we want to add two numbers that appear in scientific notation, we should expand them to ordinary decimal form if possible, and then add those quantities. Then we can convert the result back to scientific notation. Consider the following sums:

$$(3.045 \times 10^5) + (6.853 \times 10^6)$$
$$= 304{,}500 + 6{,}853{,}000$$
$$= 7{,}157{,}500$$
$$= 7.1575 \times 10^6$$

$$(3.045 \times 10^{-4}) + (6.853 \times 10^{-7})$$
$$= 0.0003045 + 0.0000006853$$
$$= 0.0003051853$$
$$= 3.051853 \times 10^{-4}$$

$$(3.045 \times 10^5) + (6.853 \times 10^{-7})$$
$$= 304{,}500 + 0.0000006853$$
$$= 304{,}500.0000006853$$
$$= 3.045000000006853 \times 10^5$$

Subtraction

Subtraction follows the same basic rules as addition. Consider the differences between the foregoing pairs of quantities:

$$(3.045 \times 10^5) - (6.853 \times 10^6)$$
$$= 304{,}500 - 6{,}853{,}000$$
$$= -6{,}548{,}500$$
$$= -6.548500 \times 10^6$$

$$(3.045 \times 10^{-4}) - (6.853 \times 10^{-7})$$
$$= 0.0003045 - 0.0000006853$$
$$= 0.0003038147$$
$$= 3.038147 \times 10^{-4}$$

$$(3.045 \times 10^5) - (6.853 \times 10^{-7})$$
$$= 304{,}500 - 0.0000006853$$
$$= 304{,}499.9999993147$$
$$= 3.044999999993147 \times 10^5$$

TIP *If two numbers have absolute values that differ by many orders of magnitude, the quantity with the smaller absolute value (the one closer to 0) vanishes into insignificance, and we can ignore it in most practical situations.*

Multiplication

When we want to multiply two numbers in power-of-10 notation, we multiply the decimal numbers (to the left of the multiplication symbol) by each other. Then we *add* the powers of 10. Finally, we convert the product to standard form. Following are three examples, using the same number pairs as before:

$$(3.045 \times 10^5) \times (6.853 \times 10^6)$$
$$= (3.045 \times 6.853) \times (10^5 \times 10^6)$$
$$= 20.867385 \times 10^{(5+6)}$$
$$= 20.867385 \times 10^{11}$$
$$= 2.0867385 \times 10^{12}$$

$$(3.045 \times 10^{-4}) \times (6.853 \times 10^{-7})$$
$$= (3.045 \times 6.853) \times (10^{-4} \times 10^{-7})$$
$$= 20.867385 \times 10^{(-4-7)}$$
$$= 20.867385 \times 10^{-11}$$
$$= 2.0867385 \times 10^{-10}$$

$$(3.045 \times 10^5) \times (6.853 \times 10^{-7})$$
$$= (3.045 \times 6.853) \times (10^5 \times 10^{-7})$$
$$= 20.867385 \times 10^{(5-7)}$$
$$= 20.867385 \times 10^{-2}$$
$$= 2.0867385 \times 10^{-1}$$
$$= 0.20867385$$

We should write out the last number in plain decimal form because the exponent lies between −3 and 3 inclusive.

Division

When we want to divide one number by another in power-of-10 notation, we take the quotient of the decimal numbers (to the left of the multiplication symbol). Then we *subtract* the powers of 10. Finally, we convert the result to standard form. To illustrate how the process works, let's calculate quotients with the same three number pairs that we've been using:

$$(3.045 \times 10^5)/(6.853 \times 10^6)$$
$$= (3.045/6.853) \times (10^5/10^6)$$
$$\approx 0.444331 \times 10^{(5-6)}$$
$$\approx 0.444331 \times 10^{-1}$$
$$\approx 0.0444331$$

$$(3.045 \times 10^{-4})/(6.853 \times 10^{-7})$$
$$= (3.045/6.853) \times (10^{-4}/10^{-7})$$
$$\approx 0.444331 \times 10^{[-4-(-7)]}$$
$$\approx 0.444331 \times 10^3$$
$$\approx 4.44331 \times 10^2$$
$$\approx 444.331$$

$$(3.045 \times 10^5)/(6.853 \times 10^{-7})$$
$$= (3.045/6.853) \times (10^5/10^{-7})$$
$$\approx 0.444331 \times 10^{[5-(-7)]}$$
$$\approx 0.444331 \times 10^{12}$$
$$\approx 4.44331 \times 10^{11}$$

Still Struggling

Note the wavy equals signs ($\approx$) in the above equations. This symbol stands for the words "approximately equals." The quotients here don't divide out neatly to produce resultants with a reasonable numbers of digits. You might naturally ask, "How many digits is reasonable?" The answer lies in the method scientists use to determine so-called *significant figures*. You'll learn about that technique in a little while!

Exponentiation

When we want to raise a quantity to a power in scientific notation, we must raise both the coefficient and the power of 10 to that power, and then multiply the exponents by each other. Consider this example:

$$(4.33 \times 10^5)^3$$
$$= 4.33^3 \times (10^5)^3$$
$$= 81.182737 \times 10^{(5 \times 3)}$$
$$= 81.182737 \times 10^{15}$$
$$= 8.1182737 \times 10^{16}$$

Now let's look at a situation with a negative exponent:

$$(5.27 \times 10^{-4})^2$$
$$= 5.27^2 \times (10^{-4})^2$$
$$= 27.7729 \times 10^{(-4 \times 2)}$$
$$= 27.7729 \times 10^{-8}$$
$$= 2.77729 \times 10^{-7}$$

Taking Roots

To find the root of a number in power-of-10 notation, we can regard the root as a fractional exponent. The square root is the same as the 1/2 power, the cube root is the same as the 1/3 power, the fourth root is the same as the 1/4 power, and so on. Once we convert the root to a fractional exponent, we can multiply things out in the same way as we would do with whole-number powers. Here's an example:

$$(5.27 \times 10^{-4})^{1/2}$$
$$= (5.27)^{1/2} \times (10^{-4})^{1/2}$$
$$\approx 2.2956 \times 10^{[-4 \times (1/2)]}$$
$$\approx 2.2956 \times 10^{-2}$$
$$\approx 0.02956$$

Once again, note the wavy equals signs! The square root of 5.27 turns out to be an irrational number, so we must approximate its decimal expansion. We can't write an irrational number out in full as a decimal expression. In order to do that, we'd have to keep writing digits forever.

Approximation, Error, and Precedence

In practical trigonometry, we rarely work with exact values. We must almost always settle for an estimate, or a value with a certain amount of imprecision. We can approximate quantities in two ways: *truncation* (the easy but less accurate way) and *rounding* (a little trickier, but the more accurate method).

Truncation

The process of truncation involves simply "chopping off" all the numerals to the right of a certain point in the decimal part of an expression. Some calculators use this process to fit numbers within their displays. For example, we can truncate the number 3.830175692803 in steps as follows:

$$3.830175692803$$
$$3.83017569280$$
$$3.8301756928$$
$$3.830175692$$
$$3.83017569$$
$$3.8301756$$
$$3.830175$$
$$3.83017$$
$$3.8301$$
$$3.830$$
$$3.83$$
$$3.8$$
$$3$$

Rounding

Most scientists and engineers prefer rounding for approximation purposes. In this process, when we delete a given digit (call it r) at the right-hand extreme of an expression, we don't change the digit q to its left (which becomes the new r after we delete the original r) if $0 \leq r \leq 4$. However, if $5 \leq r \leq 9$, then we increase q by 1. High-end scientific calculators use rounding, not truncation, when they need to approximate a quantity. We can round the number 3.830175692803 in steps as follows:

$$3.830175692803$$
$$3.83017569280$$
$$3.8301756928$$
$$3.830175693$$
$$3.83017569$$
$$3.8301757$$
$$3.830176$$
$$3.83018$$
$$3.8302$$
$$3.830$$
$$3.83$$
$$3.8$$
$$4$$

Error

Whenever we measure a physical quantity, we must expect some inexactness. Errors occur because of imperfections in the instruments, and sometimes because of human observational limitations.

Suppose that x_a represents the *actual* value of a quantity that we want to measure. Let x_m represent the *measured* value of that quantity, in the same units as x_a. We define the *absolute error* D_a (in the same units as x_a) using the equation

$$D_a = x_m - x_a$$

The *proportional error* D_p equals the absolute error divided by the actual value of the quantity. We have

$$D_p = (x_m - x_a)/x_a$$

The *proportional error percentage* $D_\%$ equals 100 times D_p. Mathematically,

$$D_\% = 100(x_m - x_a)/x_a$$

Error values and percentages turn out positive if $x_m > x_a$, and negative if $x_m < x_a$. If the measured value is too large, we have a positive error; if the measured value is too small, we have a negative error. Sometimes the possible error or uncertainty in a situation is expressed as "plus or minus" a certain number of units or percent. We indicate this state of affairs by writing a *plus-or-minus sign* (±) before the error figure.

Still Struggling

The denominators in the foregoing equations contain x_a, the actual value of a physical quantity—which we can never know exactly because we can't make perfect measurements. How can we calculate the extent of an error based on formulas containing a quantity that's subject to that very error? It's a classical conundrum! Some scientists get around this trouble by deriving a theoretical or "ideal" value of x_a from what they already know about the phenomenon, and then comparing an "official" observed value to the theoretical value. Sometimes the "official" observation equals the average of numerous measurements, each with its own value (say x_{m1}, x_{m2}, x_{m3}, and so on). As the number of measurements increases, scientists can get an increasingly believable "official" observation.

Precedence

Over the years, mathematicians, scientists, and engineers have reached a consensus on the priority or *precedence* for arithmetic operations that occur together in complicated expressions and equations. This convention prevents confusion and ambiguity. When various operations such as addition, subtraction, multiplication, division, and exponentiation appear in an expression, and if you need to simplify that expression, you can follow a sixfold sequence as if you're a computer executing a program:

1. Simplify all expressions within parentheses ( ), working your way from the inside out in each case. Use rules 4 through 6, in that order, inside each parenthetical expression, to simplify its contents.

2. Simplify all expressions within square brackets [ ], working your way from the inside out in each case. Use rules 4 through 6, in that order, inside each square-bracketed expression, to simplify its contents.

3. Simplify all expressions within curly braces { }, working your way from the inside out in each case. Use rules 4 through 6, in that order, inside each curly-braced expression, to simplify its contents.

4. Perform all exponential operations (powers), working your way through the entire expression from left to right.

5. Perform all products (multiplications) and quotients (divisions), working your way through the entire expression from left to right.

6. Perform all sums (additions) and differences (subtractions), working your way through the entire expression from left to right.

TIP *If you think that the above "legal document" is arcane, I can only say, "I agree, it's a monster, but it's a well-designed monster." Fortunately, you'll rarely encounter any expression so complicated that it forces you use all the power that this set of rules has. (If you ever do need it, though, you're armed with it now!)*

Let's simplify two expressions according to the rules of precedence. The order of the numerals and operations is identical in every case, but the groupings differ.

$$[(2 + 3)(-3 - 1)^2]^2$$
$$= [5 \times (-4)^2]^2$$
$$= (5 \times 16)^2$$
$$= 80^2$$
$$= 6400$$

$$\{[2 + 3 \times (-3) - 1]^2\}^2$$
$$= [(2 + (-9) - 1)^2]^2$$
$$= (-8^2)^2$$
$$= 64^2$$
$$= 4096$$

Suppose that you see a complicated expression that contains no grouping symbols at all. You can resolve it, no matter how "messy" it might be, as long as you carefully follow the rules of precedence. Consider

$$z = -3x^3 + 4x^2y - 12xy^2 - 5y^3$$

If you want to portray this equation with parentheses, brackets, and braces to emphasize the rules of precedence, you should write

$$z = [-3(x^3)] + \{4[(x^2)y]\} - \{12[x(y^2)]\} - [5(y^3)]$$

The rules allow you to do without the parentheses, brackets, and braces, but only as long as you're sure that you understand those rules.

TIP *Some mathematicians see elegance in minimizing the number of parentheses, brackets, and braces in mathematical expressions. However, extra grouping symbols won't do any harm if you place them correctly. You'll do better to put in a few unnecessary markings than to risk having someone interpret an expression the wrong way.*

 **PROBLEM 8-3**

Consider two vectors in mathematician's polar coordinates (MPC), as follows:

$$\mathbf{a} = (\theta_a, r_a)$$
$$= [0.000°, (3.566 \times 10^{13})]$$

and

$$\mathbf{b} = (\theta_b, r_b)$$
$$= [54.000°, (1.234 \times 10^7)]$$

Find the dot product $\mathbf{a} \bullet \mathbf{b}$, accurate to three decimal places by rounding.

 **SOLUTION**

We can find the dot product of two MPC vectors by multiplying their lengths, and then multiplying that product by the cosine of the angle between the vectors. First, let's multiply the lengths r_a and r_b:

$$r_a r_b = (3.566 \times 10^{13}) \times (1.234 \times 10^7)$$
$$= 3.566 \times 1.234 \times 10^{(13+7)}$$
$$= 4.400444 \times 10^{20}$$

TIP *In the foregoing calculation, we've expanded the product to more decimal places than the problem actually calls for. We'll round off our final answer at the very end of the calculation process. If we round anything off in the early or intermediate stages of a calculation, uncertainties in the individual quantities sometimes "conspire together" to produce a disproportionate error in the final result. Scientists and engineers refer to this phenomenon as* cumulative rounding error.

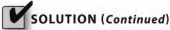

SOLUTION (*Continued*)

The angle θ between the vectors, as stated in Problem 8-3, equals $\theta_a - \theta_b =$ 54.000° − 0.000° = 54.000°. Its cosine, found using a calculator, equals 0.587785, accurate to six decimal places. Once again, we include some extra digits to prevent cumulative rounding error. Multiplying out and rounding off our final answer to three decimal places, we get

$$r_a r_b \cos \theta \approx 4.400444 \times 10^{20} \times 0.587785$$
$$\approx (4.400444 \times 0.587785) \times 10^{20}$$
$$\approx 2.587 \times 10^{20}$$

PROBLEM 8 - 4

Suppose that we encounter the same two vectors in MPC as those in the previous problem:

$$\mathbf{a} = (\theta_a, r_a)$$
$$= [0.000°, (3.566 \times 10^{13})]$$

and

$$\mathbf{b} = (\theta_b, r_b)$$
$$= [54.000°, (1.234 \times 10^7)]$$

Find the cross product $\mathbf{a} \times \mathbf{b}$, accurate to three decimal places by truncation.

SOLUTION

First, we should make certain that we know the direction in which the cross product vector points. Remember the right-hand rule. Because the angle between vectors $\mathbf{a}$ and $\mathbf{b}$ is positive but less than 180°, the vector $\mathbf{a} \times \mathbf{b}$ points up out of the page, straight toward us. We can also envision this situation by imagining that the axis representing $\theta = 0$ in the MPC plane points due east, and we express the angles going counterclockwise as viewed from some point high above the plane containing both vectors. In that case, the vector $\mathbf{a} \times \mathbf{b}$ points toward the zenith.

We can calculate the magnitude of $\mathbf{a} \times \mathbf{b}$ by multiplying the vector lengths, and then multiplying that product by the sine of the angle between the vectors. First, let's multiply their lengths, r_a and r_b:

$$r_a r_b = (3.566 \times 10^{13}) \times (1.234 \times 10^7)$$
$$= 3.566 \times 1.234 \times 10^{(13+7)}$$
$$= 4.400444 \times 10^{20}$$

The angle θ between the vectors equals 54.000°, exactly as it was in Problem 8-3. Its sine, found by using a calculator, equals 0.809016 (truncated to six decimal places). Multiplying out and truncating to three decimal places, we get

$$r_a r_b \sin \theta \approx 4.400444 \times 10^{20} \times 0.809016$$
$$\approx (4.400444 \times 0.809016) \times 10^{20}$$
$$\approx 3.560 \times 10^{20}$$

Significant Figures

When we do multiplication or division using power-of-10 notation, the number of significant figures (also called *significant digits*) in the result can't "legally" exceed the number of significant figures in the least-exact expression.

Consider the two numbers $x = 2.453 \times 10^4$ and $y = 7.2 \times 10^7$. The following statement holds true in pure arithmetic:

$$xy = 2.453 \times 10^4 \times 7.2 \times 10^7$$
$$= 2.453 \times 7.2 \times 10^{(4+7)}$$
$$= 17.6616 \times 10^{11}$$
$$= 1.76616 \times 10^{12}$$

If x and y represent measured quantities, as they would in experimental science or engineering, the above statement needs qualification.

How Accurate Can You Get?

When you see a product or quotient in scientific notation, count the number of single-digit numerals in the decimal portions of each quantity. Then identify the quantity with the smallest number of digits, and count the number of single-digit numerals in it. That's the maximum number of significant figures that you can "legally" claim in your final answer.

In the above example, there are four single digits in the decimal part of x, and two single digits in the decimal part of y. Therefore, you must round off the answer, which appears to contain six significant figures, to only two significant figures. (Use rounding, not truncation!) You can conclude that

$$xy = 2.453 \times 10^4 \times 7.2 \times 10^7$$
$$\approx 1.8 \times 10^{12}$$

No More Wavy Equals Signs!

In science and engineering, approximation is the rule, not the exception. If you want to maintain absolute rigor, you should write wavy equals signs whenever you round off any quantity, or whenever you make any observation. However, the task of scrawling wavy symbols can get tiresome in a hurry. Most scientists and engineers use ordinary equals signs, even when everybody knows that an approximation or error exists in the expression of a quantity.

Suppose that you want to find the quotient x/y in the above situation, instead of the product xy. You can proceed as follows, using straight equals signs all the way through:

$$x/y = (2.453 \times 10^4)/(7.2 \times 10^7)$$

$$= (2.453/7.2) \times 10^{(4-7)}$$

$$= 0.3406944444 \ldots \times 10^{-3}$$

$$= 3.406944444 \ldots \times 10^{-4}$$

$$= 3.4 \times 10^{-4}$$

What about Ciphers?

When you make a calculation, you'll sometimes get an answer that lands on a neat, seemingly whole-number value. Consider $x = 1.41421$ and $y = 1.41422$. Both of these quantities have six significant figures. The product, taking significant figures into account and rounding off at the end of the process, is

$$xy = 1.41421 \times 1.41422$$

$$= 2.0000040662$$

$$= 2.00000$$

Upon casual examination, this result appears exactly equal to 2. In pure mathematics, 2.00000 does indeed represent precisely the same quantity as 2 (or, for that matter, 2.0 or 2.00 or 2.0000000000000000!). But in physics or engineering, things aren't quite that straightforward. The five ciphers in the numeral 2.00000 indicate how near to the exact number 2 we believe the resultant to be. We know that the answer lies very close to a mathematician's pure idea of the number 2, but there's an inherent uncertainty of up to ± 0.000005. If we chop off the ciphers and say simply that $xy = 2$, we allow for an uncertainty of up to ± 0.5. We're entitled to a lot more precision than that, even though we can't claim that xy equals precisely 2.

TIP   *When we claim a certain number of significant figures, we must give ciphers just as much consideration as we give other digits. Ancient mathematicians invented the numeral 0 as a mere place-holder in arithmetic, but modern scientists and engineers have given it some additional "responsibilities."*

In Addition and Subtraction

When we add or subtract measured or approximate quantities, determining the number of significant figures often involves some subjective judgment. The best procedure is to expand all the values out to their plain decimal form (if possible), carry out the calculation as a pure mathematician would do, and then, at the end of the process, decide how many significant figures we can reasonably claim.

In some cases, the outcome of determining significant figures in a sum or difference resembles the result of multiplication or division. Consider the sum $x + y$, where $x = 3.778800 \times 10^{-6}$ and $y = 9.22 \times 10^{-7}$. First, we write out the numbers in "plain old decimal" form as

$$x = 0.000003778800$$

and

$$y = 0.000000922$$

Then we add, getting

$$x + y = 0.0000047008$$
$$= 4.7008 \times 10^{-6}$$
$$= 4.70 \times 10^{-6}$$

In other instances, one of the values in a sum or difference is insignificant with respect to the other. Let's say that $x = 3.778800 \times 10^4$, while $y = 9.22 \times 10^{-7}$. The "plain old decimal" expressions are

$$x = 37,788.00$$

and

$$y = 0.000000922$$

When we add these two quantities, we get

$$x + y = 37,788.000000922$$
$$= 3.7788000000922 \times 10^4$$

In this case, y is so much smaller than x that y doesn't significantly affect the value of the sum. Here, we can regard y, in relation to x or to the sum $x + y$, as the equivalent of an atom compared to an elephant! If an atom lands on an elephant, the total weight doesn't appreciably change in practical terms, nor does the presence or absence of the atom have any effect on the accuracy of the scales when we weigh the elephant. We can conclude that the "sum" simply equals the larger quantity, so

$$x + y = 3.778800 \times 10^4$$

TIP   *When a large or small number does not appear in power-of-10 notation, it's best to convert it to that form before deciding on the number of significant figures it contains. If the value begins with 0 followed by a decimal point, for example 0.0004556, it's not too difficult to figure out the number of significant digits (in this case four). But when a gigantic number is written in conventional format, the number of significant digits might not be so obvious.*

Still Struggling

You can do a couple of things to avoid falling into a quagmire of uncertainty when dealing with huge numbers. First, always write the quantities in power-of-10 notation, making the number of significant figures clear. Second, if you're in doubt about the accuracy in terms of significant figures when someone states or quotes an extreme number, ask for clarification. You're better off seeming ignorant and getting things right, than acting like you're brilliant and getting things wrong.

 PROBLEM 8 - 5

Using a calculator, find the value of sin 5.33″, rounded off to as many significant figures as you can justify.

✔ **SOLUTION**

The angle is specified to three significant figures, so that's the number of significant figures to which you can justify an answer. You're looking for the sine of 5.33 seconds of arc. First, convert this angle to degrees.

Remember that one arc second equals precisely 1/3600 of a degree. Then you get

$$\sin 5.33''$$
$$= \sin (5.33/3600)°$$
$$= \sin 0.00148°$$

Using a calculator, you conclude that

$$\sin 0.00148° = 2.58 \times 10^{-5}$$

 PROBLEM 8 - 6

Suppose that a building measures 205.55 meters high. It has a flat, level roof, with no protrusions or extensions such as railings or antennas. The sun shines down from an angle of 33.5° above the horizon (Fig. 8-2). If the building sits on flat and level terrain, how long is its shadow as measured from the near side?

 SOLUTION

You know the height of the building to an accuracy of five significant figures, but you know the angle of the sun to an accuracy of only three significant figures. Therefore, you must round off the answer to three significant figures. From the right-triangle model, you can see that the height

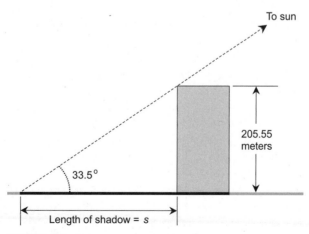

FIGURE 8-2 • Illustration for Problem 8-6.

of the building (205.55 meters) divided by the length of the shadow (the unknown, s) equals the tangent of 33.5°. You start with

$$\tan 33.5° = 205.55 / s$$

Using a calculator to determine tan 33.5° and including some extra digits to prevent cumulative rounding errors, you get

$$0.66189 = 205.55 / s$$

which solves to

$$s = 205.55 / 0.66189$$
$$= 310.55$$

That's 311 meters, rounded to three significant figures. This calculation gives you the shadow length as you would measure it from the near side of the building.

QUIZ

Refer to the text in this chapter if necessary. A good score is eight correct. Answers are in the back of the book.

1. Which of the following numerical expressions is *not* in a commonly accepted format?

   A. 3.67×10^5
   B. 36.7
   C. 0.367×10^6
   D. 367×10^3

2. If we write out the expression 3.5678E+2 in plain decimal format, it appears as

   A. 356.78.
   B. 3.5678.
   C. 35,678.
   D. 0.035678.

3. Which of the following quantities is four orders of magnitude smaller than 2.558×10^6?

   A. 2558
   B. 255.8
   C. 25.58
   D. 2.558

4. What's the product $(4.673 \times 10^6) \times (8.77 \times 10^{-11})$ taking significant figures into account?

   A. 4.673×10^6
   B. 8.77×10^{-11}
   C. 4.10×10^{-4}
   D. 5.33×10^{16}

5. What's the quotient $(4.673 \times 10^6)/(8.77 \times 10^{-11})$ taking significant figures into account?

   A. 4.673×10^6
   B. 8.77×10^{-11}
   C. 4.10×10^{-4}
   D. 5.33×10^{16}

6. What's the sum $(4.673 \times 10^6) + (8.77 \times 10^{-11})$ taking significant figures into account?

   A. 4.673×10^6
   B. 8.77×10^{-11}
   C. 4.10×10^{-4}
   D. 5.33×10^{16}

7. What's sin 359.9999°, expressed in power-of-10 notation to four significant figures? Use a calculator and then, if necessary, convert the format.

   A. 0.1745

   B. 1.745×10^4

   C. -1.745×10^{-6}

   D. 1.745×10^{-5}

8. What's tan 270.0003°, expressed in power-of-10 notation to four significant figures? Use a calculator and then, if necessary, convert the format.

   A. -0.1910

   B. 19.10

   C. -1.910×10^5

   D. 1.910×10^5

9. What's tan 269.9997°, expressed in power-of-10 notation to four significant figures? Use a calculator and then, if necessary, convert the format.

   A. -0.1910

   B. 19.10

   C. -1.910×10^5

   D. 1.910×10^5

10. What's cot 269.9997°, expressed in power-of-10 notation to four significant figures? Use a calculator and then, if necessary, convert the format. (Here's a hint: Even if your calculator lacks a cotangent function key, you can still work this problem out.)

   A. 5.236×10^{-6}

   B. -5.236×10^5

   C. 523.6

   D. -52.36

chapter 9

Surveying, Navigation, and Astronomy

Trigonometry allows us to determine distances by measuring angles. In some cases, the angles are exceedingly small, requiring complex optical apparatus. In other cases, we can measure the angles with simple devices. We can also use trigonometry to calculate angular measures (such as headings or bearings) based on known distances.

CHAPTER OBJECTIVES

In this chapter, you will

- Learn how surveyors measure distances on the surface of the earth.
- Calculate distances to objects by means of stadimetry.
- Learn how astronomers measure the distances to planets and stars.
- Participate in a "fox hunt" with radio direction finding.
- See how a vessel can "find itself" by means of radiolocation.

Terrestrial Distance Measurement

In order to measure distances using trigonometry, observers rely on a principle of classical physics: Light rays always travel in straight lines through a vacuum, and also through air having uniform density.

Parallax

Parallax allows you to judge the distances to objects, and also to perceive depth in "real-world" scenes. Figure 9-1 shows a generic example of the principle of parallax. Nearby objects appear displaced, relative to a distant background, when viewed with your left eye as compared to the view through your right eye. The extent of the displacement depends on the proportional difference between the distance to the nearby object and the distant reference scale, and also on the separation between your left eye and your right eye.

You can take advantage of parallax effects for navigation and guidance. If you're heading toward a point, say while traveling in an automobile, that point seems stationary while other objects seem to move radially outward from it. You can observe this effect while driving down a flat, straight highway. Signs, trees and other roadside objects appear to move in straight lines outward from a fixed, distant point on the road.

TIP   *Parallax simulation gives 3D video games and stereoscopic images their realism (and sometimes surrealism).*

The Base Line

If we want to use parallax for distance measurement, we must establish a *base line*: a line segment connecting two points from which we observe the distant

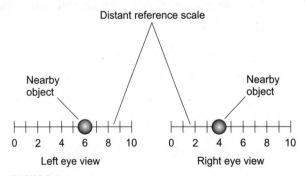

FIGURE 9-1 · Parallax displaces the apparent position of a nearby object against a distant background.

object. Let's call the observation points P and Q. If the object of interest lies at point R, then we must choose the base line such that ΔPQR constitutes a right triangle, with the right angle at one end of the base line. Therefore, line segment PQ must run perpendicular to either line segment PR or line segment QR. Figure 9-2 shows a situation for which line segment PQ runs perpendicular to line segment PR, so that $\angle RPQ$ constitutes a right angle.

TIP *At first thought, getting the base line oriented properly might seem difficult. However, because the distance of interest greatly exceeds the length of the base line in most cases, an approximate base-line orientation will usually work well enough. A hiker's compass will suffice to set the base line at a right angle to either of the two long line segments connecting the observer and the distant object.*

Accuracy

If we want to determine the distance to an object within sight, we must make sure that our two observation points (P and Q in Fig. 9-2) lie far enough apart so that a measurable difference exists in the azimuth of the object as seen from opposite ends of the base line. The absolute accuracy, in fixed units such as meters, with which we can measure the distance depends on three factors:

1. The distance to the object
2. The length of the base line
3. The precision of the angle-measuring apparatus

As the distance to the object increases, assuming that the base line length doesn't change, the absolute accuracy of the distance measurement gets worse (the error increases). As the length of the base line increases, the accuracy improves (the error decreases). As the *angular resolution*, or precision of the angle-measuring equipment, gets better, the absolute accuracy improves, if all other factors remain constant.

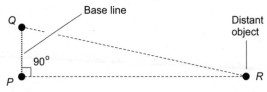

FIGURE 9-2 • We must choose a base line to measure distances using parallax.

PROBLEM 9-1

Suppose that we want to determine the line-of-sight distance to an object at the top of a mountain. Our base line measures 500.00 meters (or 0.50000 kilometers) in length. We observe an angular azimuth difference of 0.75000° from opposite ends of the base line. How many kilometers away does the object on the mountain top lie?

SOLUTION

Let's draw a diagram of this situation, even though we can't conveniently make it to scale. (We must exaggerate the length of the base line relative to the line-of-sight distance to the object.) Figure 9-3 tells the story. The base line segment PQ runs perpendicular to the long line segment PR connecting one end of the base line and the distant object. We establish the right angle $\angle RPQ$ to the best of our ability using a hiker's compass; for purposes of calculation, we can assume that $\angle RPQ = 90°$ so that $\triangle PQR$ constitutes a right triangle.

When we measure the angle θ between the long line segment QR and a ray parallel to line segment PR, we obtain 0.75000°. We can resolve the "parallel ray" by sighting to an object that lies "at an infinite distance," or by using an accurate magnetic compass.

One of the fundamental principles of plane geometry states that pairs of *alternate interior angles* formed by a *transversal* to parallel lines always have equal measure. In this example, we have line segment PR and an observation ray parallel to it, while line QR forms a transversal. Therefore, in the situation of Fig. 9-3, the two angles labeled θ have equal measure.

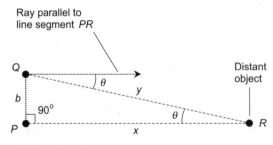

FIGURE 9-3 • Illustration for Problems 9-1 and 9-2.

We can use the triangle model for circular functions to calculate the distance to the object. Let b represent the length of the base line segment PQ, and let x represent the distance to the object along line segment PR. Then

$$\tan \theta = b/x$$

When we input the known values to this formula, we get

$$\tan 0.75000° = 0.50000/x$$

which reduces to

$$0.013090717 = 0.50000/x$$

and finally solves to

$$x = 38.195$$

The object on top of the mountain lies 38.195 kilometers away from point P.

PROBLEM 9-2

Why can't we use the length of line segment QR as the distance to the object, rather than the length of line segment PR in the above example? Or can we?

✔SOLUTION

We can indeed! For calculating the distance, observation point Q works just as well as point P does. In this situation, the base line length is only a tiny fraction of the distance that we want to determine. In Fig. 9-3, let y represent the length of line segment QR. Then, using the right-triangle model, we can state the fact that

$$\sin \theta = b/y$$

When we plug in our known values, we obtain

$$\sin 0.75000° = 0.50000/y$$

which reduces to

$$0.013089596 = 0.50000/y$$

and solves to

$$y = 38.198$$

The percentage difference between this result and the previous result is only about 3 meters out of an overall distance of roughly 38,000 meters—less than one part in 10,000, or 0.01%. In some situations, even a discrepancy that small might concern us, and we'd need to take advantage of a scheme such as *laser ranging*, in which we'd measure the distance based on a "super-accurate" knowledge of the speed of light through the atmosphere.

Stadimetry

Stadimetry allows us to measure the distance to a far-off object when we know its height or width. We determine the angular diameter of the object by careful observation, and then we calculate the distance using the right-triangle model. Stadimetry works according to the same principles as base-line distance-measurement, except that the "base line" lies at the far end of the triangle instead of at the observer's end.

Figure 9-4 shows an example of stadimetry for measuring the distance d, in meters, to a distant person. Suppose that we know the person's height h in meters. Our angle-measuring vision system (maybe a good telescope) determines the angle θ that the person's body, from head to toe, subtends in the field of view. From this information, we can calculate the distance

$$d = h / (\tan \theta)$$

In order for stadimetry to provide accurate results, the linear dimension axis (in this case the axis that depicts the person's height, h) must run perpendicular

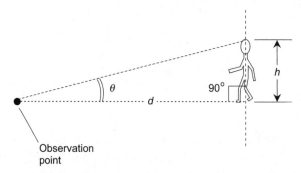

FIGURE 9-4 • Stadimetry allows us to measure the distance to an object if we know its height or width.

to a line between the observation point and one end of the object. Also, we must make sure that we express d and h in the same units, such as meters.

Interstellar Distance Measurement

Astronomers measure the distances to "nearby" stars in a manner similar to the way surveyors measure terrestrial spans. The radius of the earth's orbit around the sun serves as the base line.

The Astronomical Unit

We can express interplanetary distances in terms of the *astronomical unit* (AU), defined as the average distance of the earth from the sun, equal to approximately 1.49598×10^8 kilometers (often rounded off to 150 million kilometers). The distances to other stars and galaxies can be expressed in astronomical units, but the numbers can grow huge.

The Light Year

Astronomers use a unit called the *light year*, the distance light travels in one year, to assist in defining interstellar distances. One light year equals the distance that a ray of light travels through space in one earth year. Light rays travel approximately 3.00×10^5 kilometers in one second. A minute of time contains 60 seconds, an hour of time contains 60 minutes, a solar day contains 24 hours, and a year has about 365.25 solar days; on that basis we can calculate the fact that a light year equals roughly 9.5×10^{12} kilometers.

The nearest star to our Solar System lies slightly more than four light years away. The Milky Way, our galaxy, measures roughly 100,000 (10^5) light years across. The Andromeda galaxy, a familiar landmark for amateur and professional astronomers, lies about 2,200,000 (2.2×10^6) light years away from our Solar System. Using powerful telescopes, astronomers can peer out to distances of several billion (thousand-million) light years.

The Parsec

No one knew the true distances to the stars until the advent of the telescope, which allowed astronomers to measure extremely small angles. To determine the distances to the stars, astronomers employ *triangulation*, similar to the technique that surveyors use to measure distances on the earth's surface. Figure 9-5 shows the basic scheme, which works only for "nearby" stars.

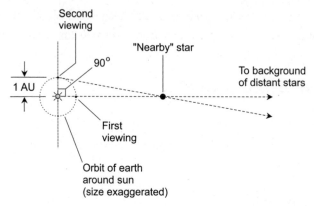

FIGURE 9-5 · We can calculate the distances to "nearby" stars by measuring the parallax resulting from the revolution of the earth around the sun.

Most stars are too far away to produce measurable parallax against a background of much more distant objects, even when we observe them from the earth at different times of the year as our planet orbits the sun. In Fig. 9-5, we exaggerate the size of the earth's orbit for clarity. The star appears in slightly different positions, relative to a background of more distant objects, from the two observation points shown. The displacement reaches its maximum when the line segment connecting the star and the sun runs perpendicular to the line segment connecting the sun with the earth.

Imagine that we look at a star whose position appears displaced by one second of arc (1″, representing 1/3600 of an angular degree) when viewed on two occasions, three months apart in time, as shown in Fig. 9-5. In that case, we define the distance between our Solar System and the star as a *parsec* (1 pc). The word "parsec" is a contraction of "parallax second". An object at a distance of 1 pc would lie approximately 3.262 light years or 2.063×10^5 AU away from us.

Sometimes units of *kiloparsecs* (kpc) and *megaparsecs* (Mpc) are used to express great distances in the universe. In this scheme,

$$1 \text{ kpc} = 1,000 \text{ pc}$$
$$= 2.063 \times 10^8 \text{ AU}$$

and

$$1 \text{ Mpc} = 10^6 \text{ pc}$$
$$= 2.063 \times 10^{11} \text{ AU}$$

TIP   *The nearest visible object outside our Solar System is the Alpha Centauri star system, 1.4 pc away. Numerous stars exist within 20 to 30 pc of our sun. The Milky Way measures roughly 30 kpc (that is, 30,000 pc) in diameter. The Andromeda galaxy is 670 kpc away. The outer limit of the observable universe, beyond which we can't see no matter how powerful our telescopes get, lies at a distance of about 3×10^9 pc, or 3000 Mpc.*

PROBLEM 9-3

Imagine that we want to determine the distance to a star. We measure the parallax relative to the background of far more distant stars, which lie "infinitely far away" by comparison. We choose the times for our observations so that the earth lies directly between the sun and the star at the time of the first measurement, and a line segment connecting the sun with the star runs perpendicular to the line segment connecting the sun with the earth at the time of the second measurement (Fig. 9-6). Suppose that the parallax thus determined equals 5.0000 seconds of arc (5.0000″). What's the distance to the star in astronomical units?

SOLUTION

We'll follow the method shown in Fig. 9-6. We should realize that the star's distance remains essentially constant throughout the earth's revolution around the sun because the star is many astronomical units away from the sun. (On an interstellar scale, ±1 AU makes practically no difference, percentage-wise.) We want to find the length of the line segment connecting the sun with the star. This line segment runs perpendicular to the line

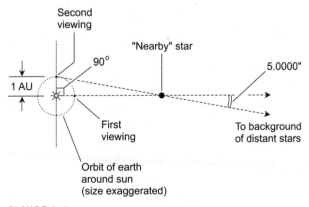

FIGURE 9-6 · Illustration for Problem 9-3.

segment connecting the earth with the sun at the time of the second observation. We, therefore, have a right triangle, and can use trigonometry to find the distance to the star in astronomical units.

The measure of the parallax in Fig. 9-6 is 5.0000″. We divide this figure by 3600 to get the number of degrees; let's call it (5/3600)° and consider it exact for now. (We'll round the answer off at the end of the calculation.) Let d represent the distance from the sun to the star in astronomical units. According to the right-triangle paradigm for the tangent function, we have

$$1/d = \tan(5/3600)°$$
$$= 2.4240684 \times 10^{-5}$$

which solves as

$$d = 41{,}252.96 \text{ AU}$$

We can round off this figure to 4.1253×10^4 AU because we're justified in going to five significant figures.

TIP   *With the help of a good scientific calculator, you can carry out the foregoing operations in sequence without having to write anything down. The display will fill up with a lot of superfluous digits, but you can (and should) round off the result when you get finished with all the arithmetic.*

Still Struggling

The parsec can give rise to confusion and gross measurement errors if we're not careful! If the distance to a star doubles, then the parallax observed between two observation points, as shown in Fig. 9-5, decreases to half of its previous amount. But the number of parsecs between us and the star doesn't go down to half. It doubles! If taken literally, the expression "parallax second" can mislead the unwary because as the number of parallax seconds goes down, the number of parsecs goes up—and vice-versa. We should always remember that the parsec is a fixed linear unit in space equal to approximately 3.262 light years or 2.063×10^5 AU, based on the distance to an object that generates a parallax of precisely 1″ as viewed from two points 1 AU apart.

Direction Finding and Radiolocation

Radio engineers and technicians sometimes use trigonometry to locate an object equipped with a wireless transmitter, based on the azimuth of that object as observed from two or more widely separated points. A well-equipped observer can also use trigonometry to locate her own position, based on the signals from two or more wireless transmitters located at widely separated, fixed points.

Radar

The acronym *radar* derives from the words *radio detection and ranging*. Radio microwaves reflect from various objects, especially if those objects contain metals or other electrical conductors. By ascertaining the direction(s) from which radio signals return, and by measuring the time it takes for a signal pulse to travel from the transmitter location to a distant object (called a *target*) and back, we can locate flying objects and evaluate some weather phenomena.

A complete radar set consists of a transmitter, a directional antenna, a receiver, and an indicator or display. The transmitter produces microwave pulses that propagate outward in a narrow beam. The waves strike objects at various distances. The greater the distance to the target, the longer the delay before the echo returns. The transmitting antenna rotates at a constant rate in a horizontal plane, facilitating observation at all azimuth bearings.

Figure 9-7 illustrates a circular radar display that employs navigator's polar coordinates (NPC). The observing station lies at the center of the display. Azimuth bearings proceed in degrees clockwise from true north, and appear in this drawing as graduations going around the perimeter of the screen. Each increment represents 10° of azimuth change. The target's distance, or range, shows up as the radial distance of the echo from the center of the screen.

Finding a "Fox"

A radio receiver, equipped with a signal-strength indicator and connected to a rotatable, directional antenna such as a *dish*, allows us to determine the direction from which wireless signals arrive. *Radio direction finding* (RDF) equipment aboard a mobile vehicle makes it possible to pinpoint the geographic location of a signal source. Sometimes hidden transmitters are employed to train people in the art of finding signal sources, or simply to serve as the subjects for a game that enthusiasts call a "fox hunt".

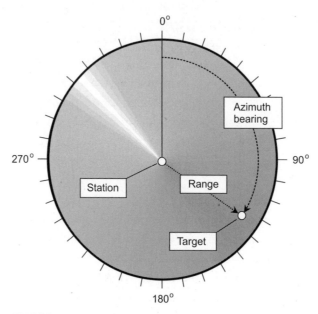

FIGURE 9-7 · A radar display shows azimuth and range in navigator's polar coordinates.

A typical RDF receiver uses a so-called *loop antenna*, comprising one or more turns of wire wound in a circle. The antenna exhibits a sharp signal minimum (called a *null*) in directions perpendicular to the geometric plane containing the loop. We turn the loop until a null occurs in the received signal strength. When we observe the null, we know that the axis of the loop coincides with a straight line that runs toward the transmitter. If we take separate readings from two or more locations separated by a sufficient distance, we can determine the transmitter's location by finding the intersection point of the azimuth bearing lines on a map or coordinate system.

Figure 9-8 shows an example of the "fox-hunting" scheme. The dots labeled *X* and *Y* indicate the locations from which we make the observations. Each of these two points forms the origin of an NPC system. The target, or "fox," appears as the shaded square. The dashed lines show the azimuth orientations of the tracking antennas at points *X* and *Y*. These lines converge on the "fox."

A "Fox" Can "Hunt" Itself!

The captain of a vessel can find his or her own position by comparing the signals from two fixed stations whose positions are known, as shown in

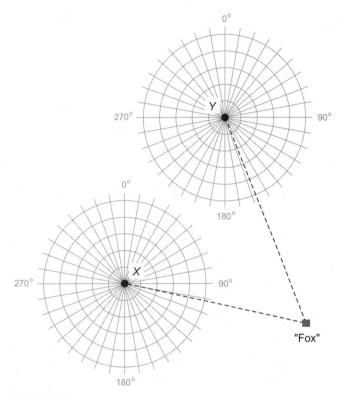

FIGURE 9-8 · We can locate a signal source using radio direction finding (RDF).

Fig. 9-9. In effect, it's a "fox hunt in reverse." A vessel, shown by the box, finds its position by taking directional readings of the signals from sources X and Y, a process called *radiolocation*. The captain can determine the vessel's direction and speed by taking two or more sets of readings separated by a certain amount of time. Computers can assist in precisely determining, and displaying, the position and velocity vectors. The term *radionavigation* refers to a process of repeated radiolocation done for the purpose of plotting or establishing a vessel's course.

Laws of Sines and Cosines

When finding the position of a target, or when trying to figure out our own location based on bearings, we should know certain rules about triangles.

The first rule is called the *law of sines*. Suppose that we define a triangle on the basis of three *vertex* points P, Q, and R. Let the lengths of the sides opposite

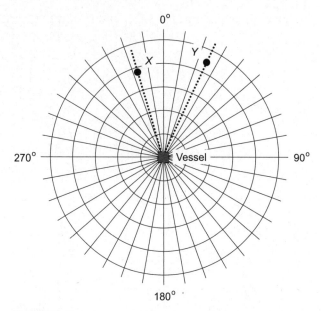

FIGURE 9-9 · A vessel can "find itself" using radiolocation.

the vertices P, Q, and R be denoted as p, q, and r, respectively (Fig. 9-10). Let the angles at vertices P, Q, and R be represented as θ_p, θ_q, and θ_r, respectively. Then

$$p/(\sin \theta_p) = q/(\sin \theta_q) = r/(\sin \theta_r)$$

The lengths of the sides of any triangle exist in a constant ratio relative to the sines of the angles opposite those sides.

The second rule is called the *law of cosines*. Suppose, once again, that we define a triangle as described above, and as portrayed in Fig. 9-10. Imagine that

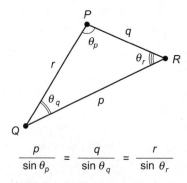

$$\frac{p}{\sin \theta_p} = \frac{q}{\sin \theta_q} = \frac{r}{\sin \theta_r}$$

FIGURE 9-10 · Illustration for the law of sines and the law of cosines.

we know the lengths of two of the sides, say p and q, and the measure of the angle θ_r between them. In that case, we can calculate the length r of the third side as

$$r = (p^2 + q^2 - 2pq \cos \theta_r)^{1/2}$$

TIP *Do you recognize the foregoing formula as a modified form of the Pythagorean theorem, which you first learned in plane geometry and which you encountered in Chap. 1 of this book?*

Global Scenarios

The above mentioned methods of direction finding and radiolocation work well over small geographic regions, within which the surface of the earth (considered as a sphere at sea level) appears nearly flat. But things get more complicated when the radio signals must travel over distances that represent an appreciable fraction of the earth's circumference. The latitude/longitude system of coordinates, or any other scheme for determining position on the surface of a sphere, involves the use of curved "lines of sight." When we use trigonometry to determine distances and angles on the surface of a sphere, we must modify the rules. We'll learn how to make such adjustments later in this course.

PROBLEM 9-4

Imagine that we use a radar set to observe an aircraft X and a missile Y, as shown in Fig. 9-11A. Aircraft X is at azimuth 240.0° and range 20.00 kilometers (km), and missile Y is at azimuth 90.0° and range 25.00 km. The two objects fly directly toward each other. Aircraft X has a ground speed of 1000 kilometers per hour (km/h), and missile Y has a ground speed of 2000 km/h. How much time will pass before the missile and the aircraft collide, assuming that neither of them changes course or speed?

SOLUTION

Let's determine the distance, in kilometers, between targets X and Y at the time of the initial observation. We have a "made-to-order job" for the law of cosines!

Consider the triangle $\triangle XSY$, formed by the aircraft X, the station S, and the missile Y, as shown in Fig. 9-11B. We know that $XS = 20.00$ km and $SY = 25.00$ km. We can also deduce that $\angle XSY = 150.0°$ (the difference between azimuth 240.0° and azimuth 90.0°). We now have a triangle with a known angle between two sides of known length. When we input the known numbers

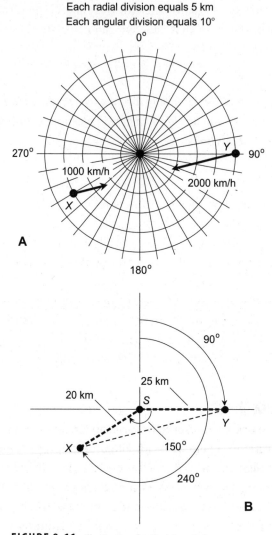

FIGURE 9-11 · Illustration for Problem 9-4.

into the formula for the law of cosines, we can calculate the distance XY between the aircraft and the missile, in kilometers, as

$$XY = [20.00^2 + 25.00^2 - (2 \times 20.00 \times 25.00 \times \cos 150.0°)]^{1/2}$$

$$= [400.0 + 625.0 - (1000 \times -0.8660)]^{1/2}$$

$$= (1025 + 866.0)^{1/2}$$

$$= 1891^{1/2}$$

$$= 43.49 \text{ km}$$

The two objects fly directly toward *each other* (not toward our radar station S), one at 1000 km/h and the other at 2000 km/h. Their mutual speed, therefore, equals the sum of their individual speeds, or 3000 km/h. If neither object changes course or speed, they'll collide after a time t_h (in hours) determined as

$$t_h = (43.49 \text{ km})/(3000 \text{ km/h})$$
$$= 0.01450 \text{ h}$$

We can obtain the time t_s (in seconds) if we multiply the above result by 3600, the number of seconds in an hour. Then we get

$$t_s = 0.01450 \times 3600$$
$$= 52.20 \text{ s}$$

TIP   *In a true-to-life scenario of the foregoing sort, a computer would perform all of the calculations in a minuscule fraction of a second, and convey the critical information to the radar operator and the aircraft pilot immediately. The pilot would then have plenty of time to take evasive action.*

PROBLEM 9-5

Imagine that you and I, a seafaring ship's captain and navigator, wish to find our location in terms of latitude and longitude to the nearest minute of arc. We use direction-finding equipment to measure the azimuth bearings of two buoys, called Buoy 1 and Buoy 2, whose latitude and longitude coordinates are known, as shown in Fig. 9-12A. We measure the azimuth bearing of Buoy 1 as 350° 0' and the azimuth bearing of Buoy 2 as 42° 30', according to the instruments aboard the vessel. (These bearings *are not* shown in Fig. 9-12, so as to avoid cluttering up the diagram too much.) How can we solve this problem without the help of our shipboard computer, which has crashed?

SOLUTION

It's one thing to plot positions on maps (as you might have seen in old war movies), but it's another task to manually work out the values to a high level of precision. Computers can do such calculations in a fraction of a second, but it'll take us a lot longer than that to grind them out manually!

We're working within a geographic region small enough so that we can consider the earth's surface perfectly flat, and we can consider the

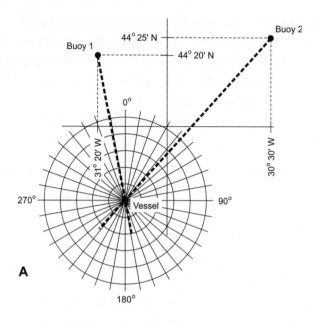

A

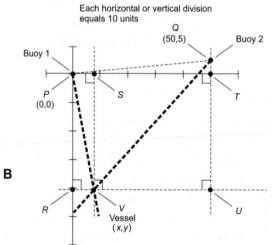

B

FIGURE 9-12 · Illustration for Problem 9-5.

lines of longitude perfectly parallel. Therefore, we can convert latitude and longitude to a rectangular coordinate grid with the origin at Buoy 1, as shown in Fig. 9-12B. Let's say that each axial division represents 10 minutes of arc.

Consider seven points P, Q, R, S, T, U, and V, representing intersections among lines and coordinate axes in Fig. 9-12B. Lines TU and SV both run perpendicular to the horizontal coordinate axis, and line VU runs perpendicular to the vertical coordinate axis. We must find either the distance *RV* or the distance *PS*. Then we can find the longitude of the vessel relative to Buoy 1. We must also determine either the distance *PR* or the distance *SV*. Then we can find the latitude of the vessel relative to Buoy 1.

As we go through the following calculations, let's consider all values exact, no matter how many significant digits they contain. We'll round the final results off to the nearest minute of arc at the end of the process. Based on the information gathered so far, we know that

$$\angle RPV = 10°$$

$$PT = 50$$

$$TQ = 5$$

We know that $\triangle TPQ$ constitutes a right triangle, so we can calculate the measure of $\angle TPQ$ on the basis of the equation

$$\tan \angle TPQ = 5/50$$

$$= 0.1$$

which solves to

$$\angle TPQ = \text{Arctan } 0.1$$

$$= 5.71059°$$

Because $\angle RPV = 10°$, we can deduce that

$$\angle VPT = 90° - 10°$$

$$= 80°$$

Because $\angle TPQ = 5.71059°$, we can deduce that

$$\angle VPQ = 80° + 5.71059°$$

$$= 85.71059°$$

We've figured out the measure of one of the interior angles of $\triangle VPQ$, an important triangle in the solution of this problem.

Now let's find $\angle PVQ$, the angle between the azimuth bearings that we've obtained. This angle equals $10° + 42° \, 30'$. Remember that $30' = 0.5°$; therefore

$$\angle PVQ = 10° + 42.5°$$

$$= 52.5°$$

From this information, we can figure out that $\angle VQP$ equals 180° minus the sum of $\angle PVQ$ and $\angle VPQ$. Inputting the numbers, we get

$$\angle VQP = 180° - (52.5° + 85.71059°)$$
$$= 180° - 138.21059°$$
$$= 41.78941°$$

We can find the distance PQ using the Pythagorean theorem because $\triangle TPQ$ is a right triangle. We know that $PT = 50$ and $TQ = 5$, and also that PQ forms the hypotenuse of the triangle. Therefore

$$PQ = (50^2 + 5^2)^{1/2}$$
$$= (2500 + 25)^{1/2}$$
$$= 2525^{1/2}$$
$$= 50.2494$$

Next, we find the distance PV by applying the law of sines to $\triangle PVQ$. We get

$$PV / (\sin \angle VQP) = PQ / (\sin \angle PVQ)$$

We can rearrange this equation and then solve it as

$$PV = PQ (\sin \angle VQP) / (\sin \angle PVQ)$$
$$= 50.2494 (\sin 41.78941°) / (\sin 52.5°)$$
$$= 50.2494 \times 0.6663947 / 0.7933533$$
$$= 42.2081$$

We now know the length of one side, along with the measures of all the interior angles, of $\triangle PRV$, which is a right triangle. We can use either $\angle RPV$ or $\angle RVP$ as the basis for finding PR and RV. Let's use $\angle RPV$, which measures 10°. Then

$$\cos 10° = PR/PV$$

We can simplify and solve this equation as

$$PR = PV \cos 10°$$
$$= 42.2081 \times 0.98481$$
$$= 41.5670$$

We also have

$$\sin 10° = RV / PV$$

which simplifies and solves as

$$RV = PV \sin 10°$$
$$= 42.2081 \times 0.17365$$
$$= 7.3294$$

The final step involves converting these units back into latitude and longitude. Keep in mind that north latitude increases from the bottom of the page to the top, but west longitude increases from the right to the left. Also remember that one angular degree contains 60 arc minutes. Let V_{lat} represent the latitude of the vessel. We can subtract the distance PR from the latitude of Buoy 1 and round off to the nearest minute of arc to get

$$V_{lat} = 44° \ 20' \ N - 41.5670'$$
$$= 43° \ 38' \ N$$

Let V_{lon} represent the longitude of the vessel. We can subtract the distance RV (which is the same as PS) from the longitude of Buoy 1 and round off to the nearest minute of arc to get

$$V_{lon} = 31° \ 20' \ W - 7.3294'$$
$$= 31° \ 13' \ W$$

Still Struggling

The process we've just completed was tedious, but it could have been worse (believe it or not). Suppose that we'd had to take the earth's curvature into account! We'd experience a real nightmare doing the calculations "by hand." Nevertheless, a computer would have an easy time of it. That's the reason why the captains of modern oceangoing vessels leave radiolocation and navigation calculations up to their computers—and always have a backup system available in case the "main brain" crashes!

QUIZ

Refer to the text in this chapter if necessary. In order to solve some of these problems, you'll need a good scientific calculator that can carry operations out to a lot of decimal places! A good score is eight correct. Answers are in the back of the book.

1. In the situation of Fig. 9-13, suppose that b equals 100 m and θ equals 20.0°, both accurate to three significant figures. What's the distance x to the nearest meter?

    A. 334 m
    B. 292 m
    C. 275 m
    D. We need more information to calculate it.

2. In the situation of Question 1 and Fig. 9-13, what's the distance y to the nearest meter?

    A. 334 m
    B. 292 m
    C. 275 m
    D. We need more information to calculate it.

3. In the situation of Fig. 9-13, suppose that b equals *exactly* 10 m and θ equals *exactly* 0.001°. What's the distance x to the nearest meter?

    A. 572,958 m
    B. 623,445 m
    C. 623,447 m
    D. We need more information to calculate it.

4. In the situation of Question 3 and Fig. 9-13, what's the distance y to the nearest meter?

    A. 572,958 m
    B. 623,445 m
    C. 623,447 m
    D. We need more information to calculate it.

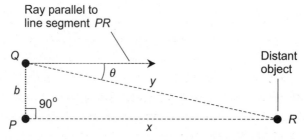

FIGURE 9-13 • Illustration for Quiz Questions 1 through 4.

5. Imagine that, as we play with the telescope at a major university's campus observatory, we find a star that lies 47 pc from our Solar System. Then we find a second, fainter star that lies 470 pc away from us in the same direction. We observe both stars from two points of view, as shown in Fig. 9-14. When measured relative to a background of galaxies whose distances we can consider "infinite," the parallax angle of the more distant star is

    A. 1/100 as large as the parallax angle of the nearer star.
    B. 1/10 as large as the parallax angle of the nearer star.
    C. 10 times as large as the parallax angle of the nearer star.
    D. 100 times as large as the parallax angle of the nearer star.

6. In the situation described by Question 5 and shown in Fig. 9-14, the more distant star lies

    A. 1/100 as far away from us as the nearer star.
    B. 1/10 as far away from us as the nearer star.
    C. 10 times as far away from us as the nearer star.
    D. 100 times as far away from us as the nearer star.

7. The law of sines tells us that

    A. as an angle approaches 90°, the ratio of the sine to the tangent approaches 0.
    B. as an angle approaches 45°, the sine approaches the cosine.
    C. as an angle approaches 0°, the sine approaches the tangent.
    D. None of the above

8. Figure 9-15 shows a triangle with two angle measures and one side length known. How long is side *x* to the nearest meter (m)? Assume that all three of the indicated values (the two angle measures and the side length) are exact.

    A. 27 m
    B. 29 m
    C. 32 m
    D. 36 m

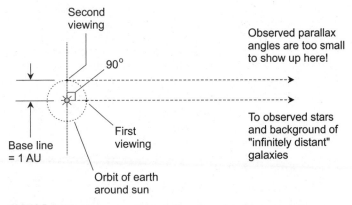

FIGURE 9-14 · Illustration for Quiz Questions 5 and 6.

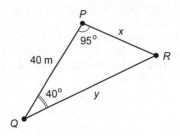

FIGURE 9-15 · Illustration for Quiz Questions 8 and 9.

9. How long is side *y* in the triangle of Fig. 9-15, to the nearest meter?

    A. 62 m
    B. 56 m
    C. 54 m
    D. 52 m

10. Figure 9-16 shows a vessel and two sounding beacons *X* and *Y*. Beacon *X* lies at azimuth 342.0° and range 38.1 km with respect to the vessel. Beacon *Y* lies at azimuth 25.2° and range 44.8 km with respect to the vessel. Based on the information given in this diagram, how far apart are the sounding beacons *X* and *Y* from each other, expressed to the nearest kilometer (km)?

    A. 29 km
    B. 31 km
    C. 33 km
    D. We need more information to calculate it.

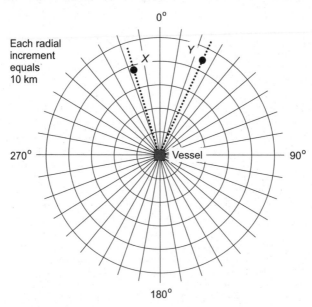

FIGURE 9-16 · Illustration for Quiz Question 10.

chapter **10**

Electrical Waves and Phase

Trigonometry plays important roles in electricity and electronics. Let's find out how the circular functions can help us analyze *alternating-current* (AC) *waves*.

CHAPTER OBJECTIVES

In this chapter, you will

- See how period and frequency relate in an AC wave.
- Define phase in terms of degrees and radians.
- Compare conventional frequency with angular frequency.
- Discover the difference between reactance and resistance.
- Learn how inductance and capacitance combine with resistance to affect the behavior of an AC wave.

Alternating Current

In electrical applications, *direct current* (DC) has a *polarity*, or direction, that stays the same for an indefinite length of time. Although the intensity of the current might vary from moment to moment, the electrons always flow in the same direction through the circuit. In AC, the polarity reverses at regular intervals. In a wire or other electrical conductor, electrons move back and forth as the current "advances" and "retreats."

Period

In a *periodic AC wave*, the type we'll be talking about in this chapter, the function of *amplitude* (intensity at a specific moment) versus time reverses at regular intervals called *periods*. One period equals the time between one repetition of the pattern or *cycle* and the next. Figure 10-1 illustrates this principle for a simple AC wave.

The period of an alternating wave can get as short as a tiny fraction of a second, or as long as thousands of centuries. Some *electromagnetic* (EM) *fields* have periods that we must express in quadrillionths of a second (units of 10^{-15} s) or even quintillionths of a second (units of 10^{-18} s). At the opposite extreme, the charged particles held captive by the *magnetic field* of the sun reverse their direction over periods measured in years, and large galaxies may have magnetic fields that reverse their polarity every few million years.

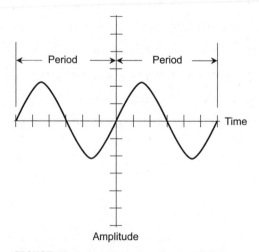

FIGURE 10-1 •We define the period of an AC wave as the time it takes for exactly one complete cycle to occur.

When we want to express the period of an AC wave, let's use the uppercase letter T. Unless otherwise specified, we express or measure T in seconds. In other words, seconds constitute the "default unit" for wave periods.

Frequency

The *frequency* of an AC wave, denoted f, equals the reciprocal of the period. If we want to get technical, we write

$$f = 1/T$$

and

$$T = 1/f$$

Prior to the 1970s, scientists and engineers expressed frequency in *cycles per second*, abbreviated cps. They expressed high frequencies in *kilocycles, megacycles,* or *gigacycles*, representing thousands, millions, or billions (thousand-millions) of cycles per second. Nowadays, the standard unit of frequency is the *hertz*, abbreviated Hz. It's really nothing more than a fancy word that means "cycles per second." Therefore, 1 Hz = 1 cps, 10 Hz = 10 cps, and so on. Higher frequencies are expressed in *kilohertz* (kHz), *megahertz* (MHz), *gigahertz* (GHz), and *terahertz* (THz), according to the following progression:

$$1 \text{ kHz} = 1000 \text{ Hz}$$
$$= 10^3 \text{ Hz}$$

$$1 \text{ MHz} = 1,000,000 \text{ Hz}$$
$$= 10^6 \text{ Hz}$$

$$1 \text{ GHz} = 1,000,000,000 \text{ Hz}$$
$$= 10^9 \text{ Hz}$$

$$1 \text{ THz} = 1,000,000,000,000 \text{ Hz}$$
$$= 10^{12} \text{ Hz}$$

The Sine Wave

In its purest form, AC has a *sine-wave* nature. The *waveform* (shape or contour of the curve) in Fig. 10-1 gives us an example of a sine wave. Any AC wave that concentrates all of its energy at a single frequency has a perfect sine-wave shape. Conversely, any perfect sine wave (sometimes called a *sinusoid*) has one, and only one, component frequency.

TIP *In an electronics lab, a wave can resemble a sinusoid so closely that it looks just like a graph of the sine function on a wave-analyzing instrument called an* oscilloscope, *when in reality, small components exist at other frequencies as well. The imperfections in a signal are often too small to see using an oscilloscope, although other, more sophisticated instruments can detect and measure them. Utility AC in the United States has an almost perfect sine-wave shape with a frequency of 60 Hz.*

Degrees of Phase

Engineers often divide an AC cycle into 360 equal increments called *degrees of phase,* (symbolized with the degree sign °). We assign 0° to the point in the cycle where the signal amplitude equals zero while increasing (usually upward in a graph). We give the same point on the next cycle the value 360°. Then, as shown in Fig. 10-2, we can identify other points:

- A quarter (1/4) of the way through the cycle equals 90°
- Half (1/2) of the way through the cycle equals 180°
- Three-quarters (3/4) of the way through the cycle equals 270°

Radians of Phase

As an alternative to degrees of phase, we can break down an AC cycle into 2π equal parts. One *radian of phase* (symbolized rad) equals $360°/(2\pi)$, or about 57.296°. Physicists often use the radian when discussing an AC cycle.

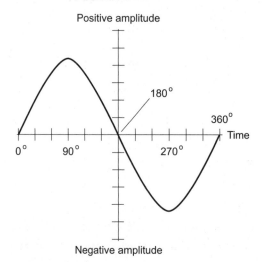

FIGURE 10-2 · One cycle of an AC waveform contains 360 degrees of phase.

Sometimes, the frequency of an AC wave is expressed in *radians per second* (rad/s), rather than in hertz (Hz). Because a complete 360° cycle contains 2π rad, the *angular frequency* of a wave, in radians per second, equals 2π times the frequency in hertz. Scientists symbolize angular frequency in equations by writing a lowercase, italicized Greek letter omega (ω).

Some engineers express angular frequency in *degrees per second* (°/s). The angular frequency in degrees per second equals 360 times the frequency in hertz, or 57.296 times the angular frequency in radians per second.

Instantaneous Amplitude

In a sine wave, the amplitude varies with time, over the course of one complete cycle, according to the sine of the number of degrees or radians measured from the start of the wave cycle (the point on the wave where the amplitude is zero and positive-going). If we represent the maximum positive amplitude, also called the *peak positive amplitude*, that a wave X attains as x_{pk+} units (such as volts or amperes), then the *instantaneous amplitude*, denoted x_i, at any instant of time is

$$x_i = x_{pk+} \sin \phi$$

where ϕ represents the number of degrees or radians between the start of the cycle and the specified point (instant) in time.

PROBLEM 10-1

What's the angular frequency of 60-Hz household AC in radians per second? Assume that the 60-Hz figure is exact. Round off the answer to three significant figures.

SOLUTION

Multiply the frequency in hertz by 2π. If we take this value as 6.2832, then we can calculate the angular frequency ω as

$$\omega = 6.2832 \times 60.000$$
$$= 377 \text{ rad/s}$$

PROBLEM 10-2

A certain wave has an angular frequency of 3.8865×10^5 rad/s. What's its frequency in kilohertz? Express the answer to three significant figures.

SOLUTION

Let's find the wave's frequency in hertz first, and then convert it to kilohertz. We must divide the angular frequency (in radians per second) by 2π, which equals approximately 6.2832. The frequency f_{Hz} is therefore

$$f_{Hz} = (3.8865 \times 10^5)/6.2832$$
$$= 6.1855 \times 10^4 \text{ Hz}$$

To obtain the frequency in kilohertz, we divide by 10^3 and then round off to three significant figures to get

$$f_{kHz} = 6.1855 \times 10^4/10^3$$
$$= 61.855 \text{ kHz}$$
$$= 61.9 \text{ kHz}$$

Phase Angle

Phase angle quantifies the displacement between two waves having identical frequencies. We usually express phase angles as values ϕ such that $0° \leq \phi < 360°$. In radians, that range equals $0 \leq \phi < 2\pi$. Once in awhile, we'll hear about phase angles over a range of $-180° < \phi \leq +180°$. In radians, that range equals $-\pi < \phi \leq +\pi$. We can define a phase angle, also called a *phase difference*, only between waves that have the same frequency. If two waves have different frequencies, their relative phase constantly changes so we can't "pin down" any definite value for it.

Waves in Phase Coincidence

Two waves exist in *phase coincidence* if and only if they have the same frequency and each cycle begins at exactly the same instant in time. When two waves coincide in phase, they appear "lined up" in a graphical display. Figure 10-3 shows this condition for two sinusoids having different amplitudes. (If the amplitudes were the same, we'd see only one wave.) The phase difference in this case equals $0°$.

TIP   *If two sine waves occur in phase coincidence, the amplitude of the resultant wave, which is also a sinusoid, equals the sum of the amplitudes of the two composite waves. The phase of the resultant coincides with the phases of both composites.*

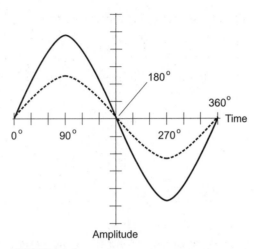

FIGURE 10-3 · Two perfect sine waves in phase coincidence. Graphically, they follow each other along.

Waves 180° out of Phase

When two pure sine waves of identical frequency begin exactly 1/2 cycle apart in time, they occur *180° out of phase* with respect to each other. Figure 10-4 illustrates a situation of this sort.

If two sine waves are identical in every respect except the fact that they exist 180° out of phase with each other, then they cancel out when we combine them because they have equal and opposite amplitudes at every instant in time.

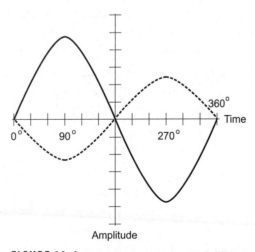

FIGURE 10-4 · Two perfect sine waves that differ in phase by 180°. Graphically, they appear ½ cycle apart.

A sine wave has the unique property that, if we shift its phase by 180°, we'll obtain the same result as we get if we *invert* the original wave ("turn it upside-down"). However, most *non-sine waves* don't behave this way. We *cannot*, in general, have confidence that a 180° phase shift is equivalent to *phase opposition* (inverting the wave).

Still Struggling

Theoretically, moving a wave earlier or later in time by 1/2 cycle doesn't change it in the same way as inverting it does. A pure sine wave represents a special case where the two actions produce the same practical result. However, with *irregular waves* or *asymmetrical* ("lopsided") *waves*, the outcomes usually differ. If you want to see these effects, graph a "lopsided" wave and then compare the results of a 180° phase shift (moving the whole set of waves to the left or right by 1/2 cycle) with the results of phase opposition (flipping the whole set of waves upside down, but not moving it to the left or right).

Leading Phase

Imagine two sinusoids, called wave X and wave Y, with identical frequencies. If X begins a fraction of a cycle earlier than Y, then we say that X *leads* Y in phase. For this situation to remain true, X must begin its cycle less than 180° before Y starts its cycle. Figure 10-5 shows X leading Y by 90°.

When one wave leads another, the phase difference exceeds 0° but stays strictly less than 180°. Engineers specify *leading phase* as a positive phase angle ϕ such that $0° < \phi < +180°$. In radians, this range equals $0 < \phi < +\pi$. If we say that X has a phase of $+\pi/2$ rad relative to Y, for example, we mean that X leads Y by $\pi/2$ rad.

Lagging Phase

Suppose that wave X starts its cycle more than 180°, but less than 360°, ahead of another wave Y that has the same frequency as X. In this situation, most people find it easier to imagine that X starts later than Y by some value between, but not including, 0° and 180°. Then we say that X *lags* Y in phase. Figure 10-6 shows wave X lagging another wave Y by 90°.

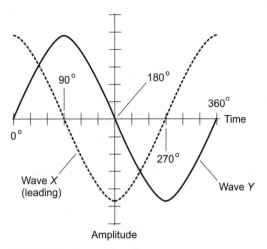

FIGURE 10-5 · Wave *X* leads wave *Y* by 90°.
Graphically, *X* appears displaced 1/4 cycle to the
left of (earlier than) *Y*.

We can portray *lagging phase* as a negative angle ϕ such that $-180° < \phi < 0°$.
In radians, this range equals $-\pi < \phi < 0$. If we say that wave *X* has a phase of
$-90°$ relative to wave *Y*, for example, we mean that *X* lags *Y* by 90°.

TIP   *When one sine wave leads or lags another wave at the same frequency by 1/4
of a cycle (90° or π/2 rad), we say that the two waves exist in* **phase quadrature.**

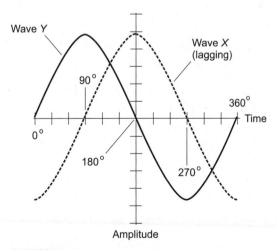

FIGURE 10-6 · Wave *X* lags wave *Y* by 90°. Graphically,
X appears displaced 1/4 cycle to the right of (later than) *Y*.

Still Struggling

If, while working out a phase problem, you find that wave X differs in phase from another wave Y by some angle ϕ that doesn't lie within the range $-180° < \phi \leq +180°$ ($-\pi < \phi \leq +\pi$ rad), you should reduce the phase difference, either positive or negative, to something that does fall within that range. You can add or subtract multiples of $360°$ (2π rad), or add or subtract whole cycles, until you get an acceptable phase difference figure. For example, suppose that someone says that wave X leads wave Y by exactly 2.75 cycles of phase. That's $2.75 \times 360°$, or $990°$. If you subtract three complete cycles ($3 \times 360°$, or $1080°$) from this figure, you end up saying that X leads Y by $-90°$, which tells you that wave X lags wave Y by $90°$.

Vectors Can Show Phase

We can portray perfect sine waves as vectors on a polar coordinate plane, where the vector length indicates the maximum positive or negative peak amplitude, and the vector direction indicates the phase. When we express sinusoids as vectors, we denote them as non-italicized boldface letters, such as vector **X** for wave X and vector **Y** for wave Y. Vectors make it easy to show phase relationships among sinusoids that all have the same frequency, but not necessarily the same amplitude. For example:

- If two sinusoids coincide in phase, then their vectors point in the same direction.
- If wave X leads Y by ϕ degrees, then we orient vector **X** at an angle of ϕ degrees *counterclockwise* from vector **Y**.
- If two sinusoids exist $180°$ out of phase with each other, then their vectors point in opposite directions.
- If wave X lags Y by ϕ degrees, then we orient vector **X** at an angle of ϕ degrees *clockwise* from vector **Y**.

Figure 10-7 shows four different phase relationships between two sinusoids X and Y. Wave X always has twice the amplitude of wave Y, so that vector **X** always has twice the length of vector **Y**.

- In Fig. 10-7A, wave X coincides in phase with wave Y.
- In Fig. 10-7B, wave X leads wave Y by $90°$ ($\pi/2$ rad).

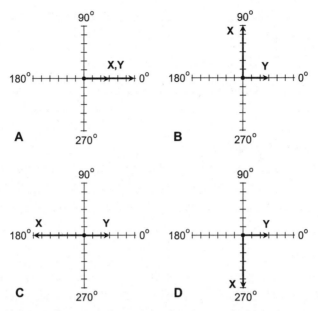

FIGURE 10-7 · Vector representations of phase difference. At A, wave X is in phase with wave Y. At B, X leads Y by 90°. At C, X and Y are 180° out of phase. At D, X lags Y by 90°.

- In Fig. 10-7C, waves X and Y are 180° out of phase with each other.
- In Fig. 10-7D, wave X lags wave Y by 90° ($\pi/2$ rad).

TIP *In all of the foregoing examples, and for all vector diagrams of sinusoids, you can imagine the vectors rotating counterclockwise around the origin (center of the graph) through one complete circle per wave cycle. You can portray the passage of time by rotating the entire plot at a constant counterclockwise rate.*

PROBLEM 10-3

Imagine three waves called X, Y, and Z. Wave X leads Y by 0.5000 rad, while Y leads Z by precisely 1/8 of a cycle. By how many degrees does wave X lead wave Z? Round off the answer to the nearest degree.

SOLUTION

Let's convert the phase angles to degrees. One radian equals about 57.296°, so

$$0.5000 \text{ rad} = 57.296° \times 0.5000$$

$$= 28.648°$$

One-eighth of a cycle equals 360.00°/ 8.0000, or 45.000°. The phase angles add up, so X leads Z by

$$28.648° + 45.000° = 73.648°$$

That's 74°, rounded off to the nearest degree.

PROBLEM 10-4

Consider three waves X, Y, and Z. Imagine that wave X leads wave Y by 0.5000 rad, while wave Y lags behind wave Z by precisely 1/8 of a cycle. By how many degrees does wave X lead or lag wave Z? Round off the answer to the nearest degree.

✔ SOLUTION

The difference in phase between X and Y is the same as in the previous problem: 28.648°. The difference between Y and Z is also the same, but in the opposite sense. Wave Y lags, rather than leads, wave Z by 45.000°. Alternatively, we can say that wave Y leads wave Z by −45.000°. Therefore, wave X leads wave Z by

$$28.648° + (−45.000°) = −16.348°$$

When we round off to the nearest degree, we get −16°. That's the extent to which X leads Z. In other words, X lags Z by 16°.

Still Struggling

In the foregoing situation, we do better to say that X lags Z by 16°, and refrain from stating that X leads Z by −16°. When we use negative leading or lagging angle values, things can get awkward. (If Jill stands behind Jack in the line at the grocery store, you don't say that Jill stands "negatively ahead of Jack," do you?)

Inductive Reactance

Electrical *resistance*—the opposition that something offers to DC—is a scalar quantity. We can always express resistance as a nonnegative value on a number line. Engineers and technicians measure it in units called *ohms*. Given a constant DC voltage, the electrical current through a device goes down as its resistance goes up. The same law holds for AC through a resistance. A component with resistance has "electrical friction." But in a coil of wire, the situation gets more complicated. A coil stores energy momentarily as a magnetic field. The coil behaves sluggishly when we drive AC through it. The coil (technically called an *inductor*) has "electrical inertia" called *inductive reactance*, which can affect AC phase.

Coils and Current

If you wind a length of wire into a coil and connect it to a source of DC, the coil grows warm as energy dissipates in the wire's resistance. If you increase the voltage, the current increases in direct proportion, and the wire gets hot. If you keep on driving up the voltage, eventually the wire will melt or vaporize!

Suppose that you change the voltage source, connected across the coil, from DC to AC, and keep the voltage low enough so that the wire doesn't melt. You vary the frequency from a few hertz (Hz) to many megahertz (MHz). The coil has a certain inductive reactance (denoted X_L), so it takes some time for current to establish itself in the coil. As you increase the AC frequency, you'll reach a point where the current can't get very well established in the coil before the AC voltage reverses. Eventually, if you keep increasing the frequency, the current will hardly begin to flow before the polarity of the voltage reverses. Under such conditions, the coil will act like it has a lot more resistance than it had for DC or for low-frequency AC. Inductive reactance, like *pure resistance*, can be expressed in ohms.

The inductive reactance of a coil can vary from zero (a short circuit) to a few ohms (for small coils or low AC frequencies) to thousands or even millions of ohms (for large coils or high AC frequencies). Like pure resistance, inductive reactance affects the current in an AC circuit. But unlike pure resistance, inductive reactance changes with frequency, influencing how the current flows with respect to the voltage.

Inductive Reactance versus Frequency

If we denote the frequency of an AC source (in hertz) as f and we denote the inductance of a coil (in units called *henrys*) as L, then we can calculate the inductive reactance X_L (in ohms) as

$$X_L = 2\pi f L$$

The value of X_L is directly proportional to f; X_L is also directly proportional to L. Figure 10-8 shows graphs of these relationships in relative form. Because the graphs appear as straight lines, we say that the relations are *linear*.

Still Struggling

Inductance stores electrical energy as a magnetic field. When a voltage appears across a coil, it takes awhile for the current to build up to its ultimate maximum. Therefore, when AC flows through a coil, the current lags the voltage in phase. The current can't keep up with the changing voltage because of "magnetic-field inertia" in the inductor. Inductive reactance and ordinary resistance combine in interesting ways. We can use trigonometry to figure out the extent to which the current lags behind the voltage in an inductance-resistance, or *RL*, electrical circuit. Let's do that now!

RL Phase Angle

Imagine a circuit that contains both "plain resistance" and inductive reactance. When the resistance is significant compared with the reactance, the AC resulting from an AC voltage lags that voltage by something less than 90° of phase, as shown in Fig. 10-9.

If the resistance R is small compared with the inductive reactance X_L, the current lag equals almost 90°; as R gets relatively larger, the lag decreases. When

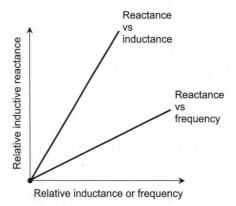

FIGURE 10-8 · Inductive reactance increases in a linear manner as the AC frequency goes up, and also as the inductance goes up.

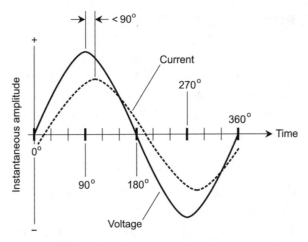

FIGURE 10-9 · An example of current that lags voltage by less than 90°, as in a circuit containing resistance and inductive reactance.

R exceeds X_L by a large factor, the so-called RL phase angle, ϕ_{RL}, is nearly zero. If the inductive reactance vanishes altogether, leaving a pure resistance, then the current and voltage follow along in phase with each other, and $\phi_{RL} = 0$.

We can determine the value of ϕ_{RL}, which represents the extent to which the current lags the voltage, using a calculator that can work with inverse trigonometric functions. The RL phase angle equals the Arctangent of the ratio of inductive reactance to resistance:

$$\phi_{RL} = \text{Arctan}\,(X_L/R)$$

PROBLEM 10-5

Find the *RL* phase angle between the AC voltage and current in an electrical circuit that has 50 ohms of resistance and 70 ohms of inductive reactance. Express your answer to the nearest whole degree.

✔ SOLUTION

Use the above formula to find ϕ_{RL}, setting $X_L = 70$ and $R = 50$. Then you can calculate

$$\phi_{RL} = \text{Arctan}\,(70/50)$$
$$= \text{Arctan}\,1.4$$
$$= 54°$$

Capacitive Reactance

Inductive reactance has a "mirror-image" counterpart in the form of *capacitive reactance*, denoted X_C. In many ways, inductive and capacitive reactance are alike. They're both forms of "electrical inertia." But in a capacitive reactance, the voltage has trouble keeping up with the current, the opposite situation from inductive reactance.

Capacitors and Current

Imagine two gigantic, flat, parallel metal plates, both of which conduct electricity very well. If we connect a source of DC, such as that provided by a large battery, to the plates with the negative battery terminal going to one plate and the positive battery terminal going to the other plate, current begins to flow immediately as the plates begin to charge up. The voltage difference between the plates starts out at zero and increases to the full DC battery voltage. The voltage buildup always takes a little bit of time because the plates take awhile to get fully charged. If the plates are small and far apart, the charging time is short. But if the plates are huge and close together, the charging time can get quite long. The plates form a *capacitor*, which stores energy in the form of an *electric field*.

Now suppose that we apply AC to the plates. We can adjust the frequency from a few hertz to many megahertz. At first, the voltage between the plates follows the voltage of the AC source as the polarity alternates. As we increase the frequency, we reach a point when the plates can't get a good charge before the polarity reverses. As we raise the frequency further, the set of plates acts more like a short circuit. Eventually, if we keep raising the frequency, current flows in and out of the plates, just as if they were connected together.

Capacitive reactance, denoted X_C, quantifies the opposition that a capacitor offers to AC. It, like inductive reactance, varies with frequency, and we can express it in ohms. By convention, engineers assign negative values to capacitive reactance. For any given capacitor, X_C increases negatively as the frequency goes down, and approaches zero from the negative side as the frequency goes up.

Capacitive Reactance versus Frequency

As previously suggested, capacitive reactance behaves like a "mirror image" of inductive reactance. In another sense, X_C constitutes an extension of X_L into negative values. If we denote the frequency of an AC source (in hertz) as f and the value of a capacitor (in units called *farads*) as C, then we can calculate X_C (in ohms) as

$$X_C = -1/(2\pi f C)$$

Capacitive reactance varies inversely with the negative of the frequency. The function of X_C versus f appears as a curve when graphed, and this curve "blows up negatively" (or, if you prefer, "blows down") as the frequency nears zero. Capacitive reactance also varies inversely with the negative of the capacitance, given a fixed frequency. Therefore, the function of X_C versus C also appears as a curve that "blows up negatively" as the capacitance approaches zero. Figure 10-10 shows relative graphs of these relations.

Still Struggling

Capacitance stores electrical energy as an electric field. When a current passes through a capacitor, it takes awhile for the voltage across the capacitor to build up to full value. Therefore, when we place an AC voltage across a capacitor, the current leads the voltage in phase. The voltage can't keep up with the changing current because of the "electric-field inertia" in the capacitor. Capacitive reactance and ordinary resistance combine in much the same way as inductive reactance combines with ordinary resistance. Let's examine that effect now!

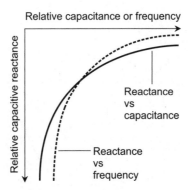

FIGURE 10-10 · Capacitive reactance varies inversely with the negative of the frequency. It also varies inversely with the negative of the capacitance, given a fixed frequency.

RC Phase Angle

When the resistance R in an electrical circuit is significant compared with the *absolute value* of the capacitive reactance, the alternating voltage resulting from an alternating current lags that current by something less than 90°. More often, an engineer will tell us that the current leads the voltage. Figure 10-11 shows an example.

If R is small compared with the absolute value of X_C, the current leads the voltage by almost 90°. As R gets relatively larger, the phase difference decreases. When R is many times greater than the absolute value of X_C, the *RC* phase angle, ϕ_{RC}, is nearly zero. If the capacitive reactance vanishes altogether, leaving a pure resistance, then the current and voltage are in phase with each other, and $\phi_{RC} = 0$.

We can determine the value of ϕ_{RC}, which represents the extent to which the current leads the voltage, using a calculator. The angle equals the Arctangent of the ratio of the absolute value of the capacitive reactance to the resistance:

$$\phi_{RC} = \text{Arctan } (|X_C|/R)$$

where the vertical lines indicate that we should take the absolute value of the enclosed quantity. Because capacitive reactance X_C is always negative or zero, we can also say that

$$\phi_{RC} = \text{Arctan } (-X_C/R)$$

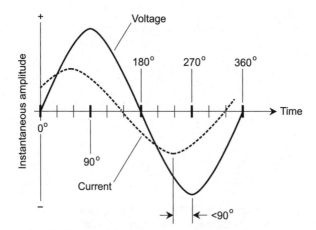

FIGURE 10-11 · An example of current that leads voltage by less than 90°, as in a circuit containing resistance and capacitive reactance.

In some texts, you'll see RC phase angles expressed as negative values themselves, indicating the extent to which the current *lags* the voltage in phase (a negative angle). In that case, the formula shows up as

$$\phi_{RC} = \text{Arctan}\ (X_C/R)$$

Still Struggling

If the notion of absolute value confuses you, remember its definition from your algebra courses. The absolute value of any number equals its "distance from 0". For real-number quantities, finding the absolute value is a trivial task. If a quantity is positive or 0, then it's the same as its absolute value. If a quantity is negative, then it's equal to the negative of its absolute value.

PROBLEM 10-6

Find the extent to which the current leads the voltage in an AC circuit that has 96.5 ohms of resistance and −21.1 ohms of capacitive reactance. Express your answer in radians to three significant figures.

✔ SOLUTION

Use the above formula to find ϕ_{RC}, setting $X_C = -21.1$ and $R = 96.5$. Then you can calculate

$$\phi_{RC} = \text{Arctan}\ (|-21.1|/96.5)$$
$$= \text{Arctan}\ (21.1/96.5)$$
$$= \text{Arctan}\ (0.21865)$$
$$= 0.215\ \text{rad}$$

QUIZ

Refer to the text in this chapter if necessary. A good score is eight correct. Answers are in the back of the book.

1. Figure 10-12 is a vector diagram showing the phase relationship between the voltage and current in a certain electronic component. Based on the information given here, the represented component contains
   A. pure inductive reactance.
   B. pure capacitive reactance.
   C. inductive reactance and resistance.
   D. capacitive reactance and resistance.

2. Based on the appearance of Fig. 10-12, the phase angle in the represented component equals
   A. $\pi/4$ rad.
   B. $\pi/2$ rad.
   C. $2\pi/3$ rad.
   D. π rad.

3. We should expect the current and voltage in an AC circuit to coincide in phase if, and only if, the circuit contains
   A. no reactance.
   B. pure capacitive reactance or pure inductive reactance.
   C. resistance and reactance.
   D. unequal amounts of capacitive and inductive reactance.

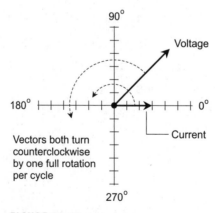

FIGURE 10-12 · Illustration for Quiz Questions 1 and 2.

4. **Based on the appearance of Fig. 10-13, the represented circuit or component contains**
   A. pure inductive reactance.
   B. pure capacitive reactance.
   C. inductive reactance and resistance.
   D. capacitive reactance and resistance.

5. **In the situation portrayed by Fig. 10-13, suppose that the absolute value of the reactance equals 100 ohms. What's the resistance?**
   A. Zero because the represented circuit contains only reactance.
   B. 327 ohms
   C. 30.6 ohms
   D. 29.2 ohms

6. **What's the period of an AC wave whose frequency equals 500 kHz?**
   A. 5.00×10^{-5} s
   B. 2.00×10^{-6} s
   C. 5.00×10^{-7} s
   D. 2.00×10^{-7} s

7. **For the wave described in Question 6, how much time would represent exactly 90° of phase?**
   A. 5.00×10^{-5} s
   B. 2.00×10^{-6} s
   C. 5.00×10^{-7} s
   D. 2.00×10^{-7} s

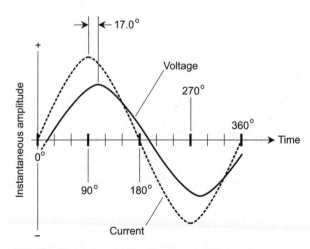

FIGURE 10-13 · Illustration for Quiz Questions 4 and 5.

8. What's the angular frequency of a wave in which 0.001000 s represents exactly π/4 rad of phase?

    A. 250π rad/s
    B. 125π rad/s
    C. 62.5 rad/s
    D. 1000 rad/s

9. Consider an RC circuit in which the current leads the voltage by 30° at a certain AC frequency. If we increase the frequency, the current will lead the voltage by

    A. 30°.
    B. more than 30°.
    C. less than 30°.
    D. no amount whatsoever because the current never leads the voltage in an RC circuit!

10. Imagine an RL circuit in which the current and voltage exist 45° out of phase at a certain AC frequency. If we double the frequency, the phase angle will

    A. remain at 45°.
    B. increase to approximately 63°.
    C. decrease to approximately 27°.
    D. change, but we need more information to calculate its exact value.

Geometrical Optics

The science of *optics* involves the effects of physical substances and objects on reflected and transmitted light. *Geometrical optics* is the mathematical analysis of everyday optical effects. Trigonometry comes in handy here!

CHAPTER OBJECTIVES

In this chapter, you will

- Observe the behavior of light rays as they reflect from flat mirrors.
- See how convex mirrors scatter light rays.
- Learn how concave mirrors focus light rays or make them parallel.
- Quantify refraction as light passes through the boundary between different media.
- Calculate refraction angles using the sine function.
- Determine critical internal reflection angles using the Arcsine function.
- Analyze the color dispersion in a glass prism.

Reflection

Any smooth surface reflects some of the light that strikes it. If the surface is perfectly flat and shiny such as the face of a pane of glass, an incoming (or *incident*) light ray reflects away at the same angle at which it hits. We've all heard the expression, "The angle of incidence equals the angle of reflection." Physicists call this principle the *law of reflection* (Fig. 11-1).

Angles

In optics, we define the angle of incidence and the angle of reflection relative to a line that runs *normal* (perpendicular) to the surface at the point where reflection takes place. In Fig. 11-1, we denote these angles as θ. They can range from 0°, when the light ray strikes at a right angle with respect to the surface, to almost 90°, a grazing angle relative to the surface.

If a reflective surface is shiny but not flat, then the law of reflection applies for any ray of light striking the surface at a specific point. In such a case, we must consider the reflection with respect to a line normal to a flat plane that passes through the point and is *locally tangent* to the surface at that point.

When parallel rays of light strike a warped, shiny surface at many different points, each individual ray obeys the law of reflection, but the reflected rays don't all emerge parallel. In some cases the emerging rays converge (come

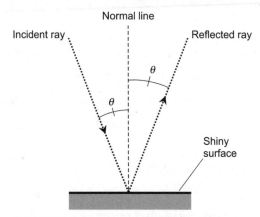

FIGURE 11-1 · When a light ray strikes a shiny surface, the angle of incidence equals the angle of reflection. In this example, we denote both angles as θ, and we express them with respect to a line normal (perpendicular) to the surface at the point of reflection.

together). In other cases they diverge (spread apart). In still other cases they come off in a haphazard fashion (at all sorts of angles).

TIP *When we say that a plane is locally tangent to a warped surface at a particular point, we mean to say that the plane doesn't intersect (touch or pass through the surface anywhere else) in the vicinity of that point. Of course, the meanings of "locally" and "vicinity" depend on the size of the region we're talking about!*

TIP *If a light beam, or a set of parallel light rays, encounters a* matte *surface (one that's not shiny, but not too rough either) such as a sheet of paper or a plaster wall, the reflected rays scatter. On a microscopic scale or for individual photons (light "particles"), the law of reflection holds. On a large scale, however, the law fails.*

A Mirrored Wall

Imagine a square room that measures five meters (5.000 m) by 5.000 m, with a mirror completely covering one of the walls. You stand near one wall (*W* as shown in Fig. 11-2A), and hold a flashlight so that its bulb lies 1.000 m away from *W* and 3.000 m away from the mirrored wall. You aim the flashlight

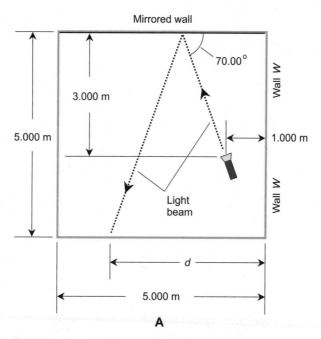

FIGURE 11-2A · A light beam reflects from a mirror and strikes the wall opposite the mirror. In this example, the angle of incidence and the angle of reflection both equal 20.00°.

horizontally at the mirrored wall, so that the center of its beam strikes the mirror at an angle of 70.00° relative to the surface. The beam reflects from the mirror and encounters the wall opposite the mirror at a certain distance *d* from wall *W*.

The light beam travels in a plane parallel to the floor and the ceiling (because, remember, you aim the flashlight horizontally). Figure 11-2B illustrates the situation in some extra detail. The center of the beam strikes the mirror at 20.00° relative to the normal. According to the law of reflection, the beam reflects from the mirror at 20.00° relative to the normal. The beam's path forms hypotenuses of two right triangles, one whose base length in meters equals *e* and whose height in meters equals 3.000, and the other whose base length in meters equals *f* and whose height in meters equals 5.000. You can calculate *e* in steps as follows:

$$\tan 20.00° = e/3.000$$

$$0.36397 = e/3.000$$

$$e = 0.36397 \times 3.000$$
$$= 1.09191 \text{ m}$$

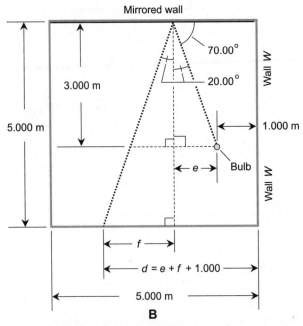

FIGURE 11-2B · Geometrical analysis of the situation shown in Figure 11-2A.

You can calculate f in a similar way:

$$\tan 20.00° = f/5.000$$

$$0.36397 = f/5.000$$

$$f = 0.36397 \times 5.000$$

$$= 1.81985 \text{ m}$$

Knowing both e and f, you can calculate d as follows:

$$d = e + f + 1.000$$

$$= 1.09191 + 1.81985 + 1.000$$

$$= 3.91176 \text{ m}$$

You should round this result off to 3.912 m because your input data is given to only four significant figures. (The extra figures in the intermediate calculations minimize the risk of cumulative rounding error.)

Virtual Image with a Flat Mirror

Suppose that you stand in front of a mirror that hangs on a wall, as shown in Fig. 11-3. You see your reflection in the mirror. If you stand at a distance d from the wall, then the image of your body, from your point of view, appears on the

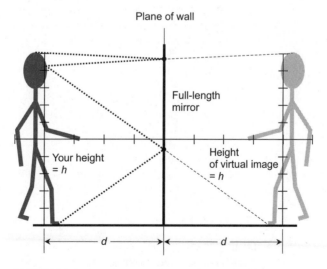

FIGURE 11-3 · When you see your reflection in a flat wall mirror, you see a virtual image.

opposite side of the wall, at the same distance d from the silvered surface of the mirror as your real body stands. When you see the image of a specific point on your body (such as the top of your head or the tip of your right shoe), you know that light rays from that point have struck the mirror, reflected from its silvered surface according to the law of reflection, and entered your eyes. In the example of Fig. 11-3, the collection of reflected rays from all the points on your body, arriving at either of your eyes, forms a *virtual image*.

TIP   *In geometrical optics, the term "virtual" refers to the fact that you can't render an image on a screen or on film.*

The Convex Mirror

A *convex mirror* reflects light so that parallel incoming rays spread out uniformly as they reflect from the surface, as shown in Fig. 11-4A. Converging incident rays, if the *angle of convergence* is just right, are *collimated* (made parallel) by a convex mirror, as illustrated in Fig. 11-4B. When you look at the reflection of a scene in a convex mirror, the objects all appear reduced in size. The field of vision is enlarged. The extent to which a convex mirror spreads light rays depends on the mirror's *radius of curvature*. As the radius of curvature decreases (the curvature gets "sharper"), the extent to which parallel incident

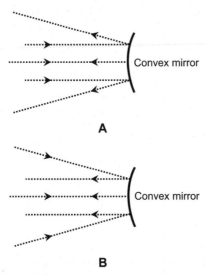

FIGURE 11-4 • At A, a convex mirror spreads parallel incident light rays. At B, the same mirror collimates converging incident light rays.

rays diverge after reflection increases. The convergence of incoming rays necessary to produce parallel reflected rays also increases.

The Concave Mirror

A *concave mirror* reflects light rays in the reverse sense from a convex mirror. When the mirror has the proper contour, and when incident rays arrive parallel to the axis of the mirror, the rays reflect so that they converge at a *focal point* (also called a *focus*) that lies on the axis, as shown in Fig. 11-5A. A *reflecting telescope* takes advantage of this fact. When we place a *point source* (an extremely tiny source) of light at the focal point, a concave mirror reflects the rays so that they emerge parallel, as shown in Fig. 11-5B. A flashlight, lantern, or spotlight works according to this principle. We call the distance of the focal point from the center of the concave mirror the *focal length*.

Some concave mirrors have spherical surfaces, but the ideal contour follows a *paraboloid*, a three-dimensional figure produced by rotating a *parabola* around its axis. When the radius of curvature is large compared with the size of the reflecting surface or when we don't need high precision, the difference between a *spherical mirror* and a *paraboloidal mirror* (often imprecisely called a *parabolic mirror*) doesn't amount to much. However, the difference does

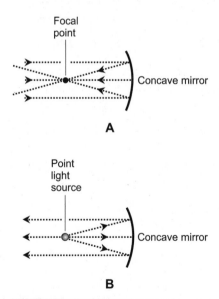

FIGURE 11-5 · At A, a concave mirror focuses parallel light rays to a point. At B, the same mirror collimates light from a source at the focal point.

matter if we want to use the mirror to build a precision optical instrument such as a reflecting telescope.

Real Image with a Concave Mirror

Imagine that you stand far away from a spherical or paraboloidal mirror, as shown in Fig. 11-6. Rays of light reflect from your body and reach the mirror surface. At each point where a ray hits the mirror, the angle of incidence equals the angle of reflection. The rays converge to form a tiny inverted image of your body near the mirror. If you place photographic film or a photosensitive screen at the location of this image, the image appears clearly. Because of this phenomenon, physicists call the focused image from a spherical or paraboloidal mirror a *real image*.

The distance of a real image from a concave mirror depends on the distance of the actual object from that same mirror. If you're "infinitely far" (a great many times the focal length) away from the mirror, then your body's real image appears at the focal point. If you're not "infinitely far" from the mirror, then the real image forms at a distance greater than the focal length. Suppose that you stand along the mirror axis at a distance r_o from the center of the mirror, where r_o significantly exceeds the focal length f, but isn't so great that you'd call it "infinity." Under these circumstances, the distance r_i of your body's real image from the mirror relates approximately to r_o and f according to the equation

$$1/f = 1/r_i + 1/r_o$$

If you want this equation to work, you must express all the distances in the same unit, such as meters or feet. The formula represents an approximation

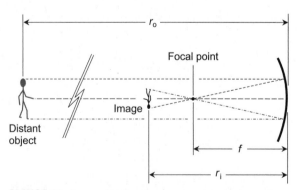

FIGURE 11-6 · A real image forms in a concave mirror when reflected rays come to a focus. The distance r_i of the image from the mirror depends on the distance r_o of the object from the mirror, relative to the focal length f.

because, in a practical situation, not every point on an observed object lies at exactly the same distance r_o from the center of the mirror. The discrepancy remains small if r_o exceeds f by a large factor. However, as r_o decreases with respect to f, the discrepancy increases for an object of fixed lateral diameter.

TIP *If you place an object at a distance less than f from a concave mirror (closer to the mirror than the focal length), a real image doesn't form because the reflected rays can't converge to a focal point.*

PROBLEM 11-1

In the situation of Fig. 11-3, your height equals h, and the height of your virtual image also equals h. If you mount the wall mirror at precisely the right distance from the floor, the mirror height only needs to equal half your body height in order for you to see your whole reflection, regardless of how far from the wall you stand. Demonstrate the reason why this rule holds true.

SOLUTION

Figure 11-7 illustrates the geometry of this scenario in detail. Suppose that your height equals h, the vertical distance from the point E between your eyes to a point H on the top of your head equals h_1, and the vertical distance

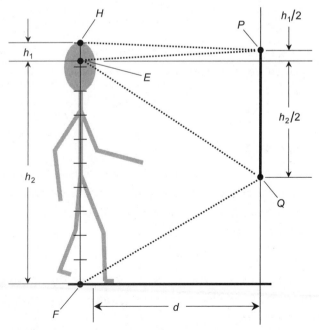

FIGURE 11-7 · Illustration for Problem 11-1.

from point E to a point F on the floor midway between your feet equals h_2. If you express all three of these distances in the same unit (such as meters or feet), then

$$h = h_1 + h_2$$

For simplicity, imagine that points H, E, and F all lie along a single vertical line. You look at the image of point H as it reflects from the mirror. Light rays reflect from point H, travel in a straight line, strike the mirror at point P, and reflect back along another straight line to point E. These three points form an isosceles triangle (let's call it $\triangle HPE$) because the distance HP equals the distance PE. Therefore, the vertical height difference between points P and E equals half the vertical height difference between points H and E. That distance equals $h_1/2$. Similarly, light rays reflect from point F, travel in a straight line, strike the mirror at point Q, and reflect back along another straight line to point E. These three points form another isosceles triangle (let's call it $\triangle FQE$) because the distances FQ and QE are the same. It follows that the vertical height difference between E and Q equals half the vertical height difference between F and E, or $h_2/2$. The vertical distance PQ between points P and Q on the mirror is therefore

$$PQ = h_1/2 + h_2/2$$
$$= (h_1 + h_2)/2$$
$$= h/2$$

If the height of the mirror equals PQ, then the mirror is tall enough to let you see your entire virtual image because $PQ = h/2$, which represents half your height.

PROBLEM 11-2

Imagine that you stand in front of a concave mirror that has a focal length of 0.200 m. You position yourself so that the mirror axis passes through your body at a point 8.00 m from the center of the mirror. How far from the center of the mirror does your real image lie?

SOLUTION

Your body's distance from the mirror is $r_o = 8.00$ m. The mirror's focal length is $f = 0.200$ m. The formula relating focal length, image distance, and object distance indicates that

$$1/f = 1/r_i + 1/r_o$$

Plugging in the known values and proceeding with the calculation step-by-step, you get

$$1/0.200 = 1/r_i + 1/8.00$$

$$5.00 = 1/r_i + 0.125$$

$$5.00 - 0.125 = 1/r_i$$

$$4.875 = 1/r_i$$

$$r_i = 1/4.875$$

$$= 0.205 \text{ m}$$

Your real image lies 0.205 m from the center of the mirror. This value represents an approximation because your height constitutes a significant fraction of your distance from the mirror.

Refraction

A clear pool looks deceptively shallow because of *refraction*, which occurs when light crosses a defined boundary from one transparent medium into another, and changes speed at the transition. In a vacuum, light rays travel at about 2.99792×10^8 meters per second (m/s). In air, light travels a little more slowly than it does in a vacuum; for lay people the difference isn't worth worrying about. In transparent liquid or solid media (such as water, glass, quartz, or diamond), however, light travels at a considerably slower speed than it does in a vacuum or in the air.

Index of Refraction

Scientists define a medium's *refractive index* (or *index of refraction*) as the ratio of the speed of light in a vacuum to the speed of light in that medium. If c represents the speed of light in a vacuum and c_m represents the speed of light in some transparent medium called M, then we define the index of refraction r_m for the medium M as

$$r_m = c/c_m$$

We must use the same speed units, such as meters per second (m/s) or kilometers per second (km/s), when expressing c and c_m. According to this definition, the index of refraction of a transparent material always exceeds 1 because c_m is always less than c.

TIP *Table 11-1 lists the approximate refractive indices for various transparent materials. These values apply to light near the middle of the visible spectrum at a wavelength of approximately 600 nanometers (nm), where 1 nm = 10^{-9} m.*

Still Struggling

As the index of refraction for a transparent substance increases, so does the extent to which a ray of light deflects when it strikes the boundary between that substance and the air at some angle other than the normal (perpendicular).

TABLE 11-1 Approximate indices of refraction for some common materials. A vacuum has a refractive index of exactly 1.

Substance	Index of Refraction
Air	1.00
Alcohol, ethyl	1.36
Amber	1.55
Calcite	1.66
Cubic zirconia	2.16
Diamond	2.42
Glass, acrylic	1.49
Glass, borosilicate	1.47
Glass, crown	1.52
Glass, flint	1.61
Glycerin	1.47
Quartz	1.54
Ruby	1.76
Salt (sodium chloride), crystal	1.50
Water, fresh, as ice	1.31
Water, fresh, as liquid	1.33

Light Rays at a Boundary

Figure 11-8 illustrates how refraction works. In Fig. 11-8A, the refractive index of the first medium (below) exceeds the refractive index of the second medium (above). A ray striking the boundary at 0° relative to the normal passes through the boundary without changing direction. Any ray that hits at some other angle gets bent as it crosses the boundary. As the angle of incidence (the angle between the incoming ray and the normal line) increases, so does the angle by which the beam deflects at the boundary, and the *angle of refraction* (the angle between the emerging ray and the normal line) exceeds the angle of incidence. When the angle of incidence reaches a certain *critical angle*, then the light ray reflects back into the first medium when it strikes the boundary, a phenomenon known as *total internal reflection*.

Total internal reflection occurs only when the second medium has a *lower* index of refraction than the first medium. If a ray of light passes from a substance

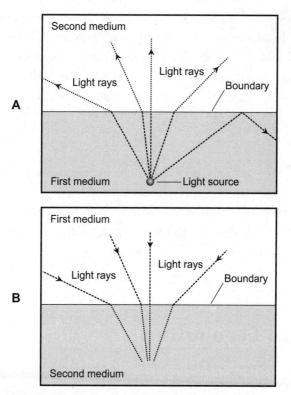

FIGURE 11-8 · At A, light rays strike a boundary where the refractive index decreases. At B, light rays strike a boundary where the refractive index increases.

having a certain refractive index into something with an *equal or higher* refractive index, total internal reflection never occurs, regardless of the incidence angle, so no critical angle exists. Nevertheless, in most situations, *some* of the light will reflect at the boundary for any angle of incidence, just as some of the incoming light reflects from the outside surface of a window on a bright day.

In the atmosphere, the speed of light varies slightly, depending on the density of the air in a particular location. Warm air has lower density than cool air at the same pressure and humidity. Because of this difference in density, warm air has a lower refractive index than cool air at similar altitudes above the earth's surface. The difference in the refractive index of warm air compared with cooler air can produce total internal reflection if a sharp boundary exists between the air masses. This effect explains why, on warm days, you sometimes see "false ponds" over the surfaces of "blacktop" highways or over stretches of hot earth.

Now consider what happens when the light rays go the other way, as shown in Fig. 11-8B. As the angle of incidence increases, so does the angle by which the beam deflects at the boundary, but the angle of refraction is less than the angle of incidence. This phenomenon causes distortion of landscape images when viewed from underwater. If you've ever gone SCUBA diving, you've witnessed it. The sky, trees, hills, buildings, people, and everything else above the horizon appear within a circle of light that distorts the scene. If you look upward toward the water surface at a near-grazing angle, you see an inverted reflection of the bottom of the body of water, or of other objects in the water with you, just as if the water surface were a mirror. That's total internal reflection!

If the refracting boundary isn't flat, the principles shown by Fig. 11-8 apply for each ray of light striking the boundary at any specific point. We define or measure the extent of the refraction with respect to a line that runs normal to a flat plane passing through the point, when the plane is tangent to the boundary at that point.

Still Struggling

When many "infinitely thin" parallel rays of light strike a curved or irregular refractive boundary at many points, each *individual* ray obeys the principle of Fig. 11-8. A "broad beam" of light gets focused or scattered, however, because individual rays refract to different extents.

Snell's law

When a ray of light encounters a boundary between two substances having different indices of refraction, we can calculate the angle by which the ray deflects according to an equation called *Snell's law*.

Figure 11-9 illustrates the principle of Snell's law for a ray of light passing from a medium with a certain refractive index to a medium having a relatively higher refractive index. Consider a flat boundary B between two media M_r and M_s, whose indices of refraction equal r and s, respectively. In this case, $r < s$.

Now consider a line N passing through point P on plane B, such that N runs normal to plane B at point P. Imagine a light ray R traveling through M_r that strikes plane B at point P. Let θ represent the angle that R subtends relative to the normal line N. Let S represent the ray of light that emerges from P into M_s. Let ϕ represent the angle that S subtends relative to the normal line N. In this situation, line N, ray R, and ray S all lie in the same plane, and $\phi < \theta$. The angles θ and ϕ will have equal measure *if and only if* the angle of incidence equals 0°. Snell's law tells us that, in general, the angles θ and ϕ relate according to the equation

$$\sin \phi / \sin \theta = r/s$$

which we can also state as

$$s \sin \phi = r \sin \theta$$

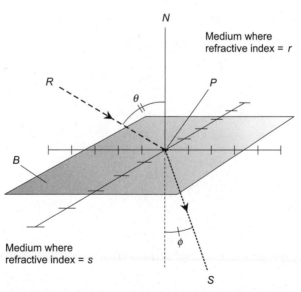

FIGURE 11-9 • Analysis of a light ray's behavior as it strikes a boundary where the index of refraction increases.

Figure 11-10 shows the situation for a ray of light passing from a medium with a certain refractive index to a medium having a relatively lower refractive index. Again, let B represent a flat boundary between two media M_r and M_s, whose absolute indices of refraction are r and s, respectively. In this case, $r > s$. Let N, B, P, R, S, θ, and ϕ represent the same things as they did in the previous example. In this case, line N, ray R, and ray S all lie in the same plane, and $\phi > \theta$. As in the previous example (Fig. 11-9), the angles θ and ϕ will have equal measure *if and only if* the angle of incidence equals 0°. The equation relating the angles and the refractive indices also remains the same as in the previous scenario, so

$$\sin \phi / \sin \theta = r/s$$

which we can also state as

$$s \sin \phi = r \sin \theta$$

Determining the Critical Angle

In the situation shown by Fig. 11-10, the light ray passes from a medium having a relatively higher index of refraction, r, into a medium having a relatively lower index, s. As the angle of incidence θ increases, the angle of refraction ϕ approaches 90°, and ray S gets closer to the boundary plane B. When θ gets large enough, ϕ reaches 90°, and ray S lies exactly in plane B. If θ increases even more, ray R undergoes total internal reflection, and we observe no ray S at all.

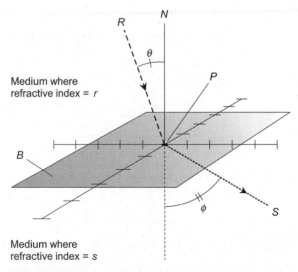

FIGURE 11-10 · Analysis of a light ray's behavior as it strikes a boundary where the index of refraction decreases.

We define the critical angle as the largest angle of incidence that ray R can subtend, relative to the normal line N, without reflecting internally. Let's call this angle θ_c. The measure of the critical angle equals the Arcsine of the ratio of s, the smaller (second medium's) index of refraction, to r, the larger (first medium's) index of refraction. We can state this rule as the equation

$$\theta_c = \text{Arcsin}\,(s/r)$$

PROBLEM 11-3

Imagine that we place a laser device beneath the surface of a liquid freshwater pond. Pure liquid water has a refractive index of approximately 1.33, while air has a refractive index of 1.00 "for all intents and purposes." Imagine that the pond has a perfectly smooth surface, without a wave or ripple. If we aim the laser beam upward so that it strikes the surface at an angle of 30.0° relative to the normal, then at what angle, also relative to the normal, will the beam emerge from the surface into the air?

SOLUTION

Envision the situation in Fig. 11-10 upside down, so that M_r represents the water and M_s represents the air. In that case, the indices of refraction are $r = 1.33$ and $s = 1.00$. The measure of angle θ equals 30.0°. The measure of angle ϕ constitutes our unknown. We can use the equation for Snell's law, plug in the numbers, and solve for ϕ in steps as follows:

$$\sin\phi/\sin\theta = r/s$$

$$\sin\phi/(\sin 30.0°) = 1.33/1.00$$

$$\sin\phi/0.500 = 1.33$$

$$\sin\phi = 1.33 \times 0.500$$
$$= 0.665$$

$$\phi = \text{Arcsin}\,0.665$$
$$= 41.7°$$

PROBLEM 11-4

What's the critical angle for light rays shining from the bottom up toward the surface of a liquid freshwater pond?

SOLUTION

Let's use the formula for determining the critical angle, and envision the scenario of Problem 11-3 where the angle of incidence, θ, can vary. We plug in the numbers to the equation for critical angle, θ_c, obtaining

$$\theta_c = \text{Arcsin } (s/r)$$
$$= \text{Arcsin } (1.00/1.33)$$
$$= \text{Arcsin } 0.752$$
$$= 48.8°$$

PROBLEM 11-5

Suppose that we place a laser device above the surface of a smooth, liquid fresh-water pool that has uniform depth everywhere. We aim the laser beam downward so that the light ray strikes the surface at an angle of 28° relative to the plane of the surface (*not* relative to a normal line). At what angle, relative to the plane of the pool bottom, will the light beam strike the bottom?

SOLUTION

Figure 11-11 illustrates this situation. The angle of incidence θ equals 90° − 28°, or 62°. We know from Table 11-1 that r, the index of refraction of the air, equals 1.00, and also that s, the index of refraction of the water, equals 1.33. We can therefore solve for the angle ϕ, relative to the normal N to the surface, at which the ray travels under the water. We proceed with the calculation, step-by-step, as follows:

$$\sin \phi / \sin \theta = r/s$$

$$\sin \phi / \sin 62° = 1.00/1.33$$

$$\sin \phi / 0.883 = 0.752$$

$$\sin \phi = 0.752 \times 0.883$$
$$= 0.664$$

$$\phi = \text{Arcsin } 0.664$$
$$= 42°$$

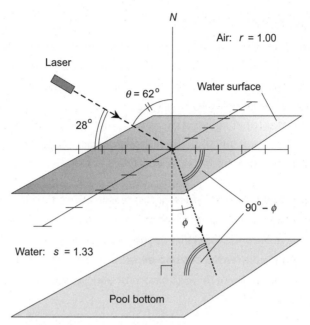

FIGURE 11-11 • Illustration for Problem 11-5.

We're justified to go to only two significant figures because that's the extent of the accuracy of the input data for the angles. The angle at which the laser travels under the water, relative to the water surface, equals 90° − 42°, or 48°. Because the pool has uniform depth, the pool bottom is flat and parallel to the water surface. We can conclude that the light beam strikes at an angle of 48° relative to the plane of the pool bottom.

Color Dispersion

In most cases, the refractive index of a transparent substance changes as we vary the wavelength of the light passing through it. Glass, and virtually any other material having a refractive index greater than 1, slows light down the most at the shortest wavelengths (blue and violet), and the least at the longest wavelengths (red and orange). This effect, which physicists call *color dispersion*, explains how a prism "splits" white light, which contains energy at many different wavelengths, into its constituent colors, each of which contains energy at a specific wavelength.

"Rainbow" Spectra

The extent to which the path of a light ray deflects in a prism depends on the refractive index of the prism material. As the index of refraction increases, so does the deflection (if all other factors hold constant). This phenomenon causes a prism to cast a *"rainbow" spectrum* (Fig. 11-12). The effect also causes the multicolored "glitter" of clear substances with high indices of refraction such as quartz and diamond. In general, as the index of refraction of a medium increases, so does the extent of the color dispersion when white light passes through it.

TIP *Dispersion is important in optics for two reasons. First, the effect allows us to make a spectrometer, a device for examining the intensity of visible light at specific wavelengths. Second, dispersion degrades the quality of white-light images viewed through simple lenses, causing multicolored borders to appear around the images of objects viewed through binoculars, telescopes, or microscopes with low-quality lenses.*

PROBLEM 11-6

Imagine that a ray of white light, shining horizontally, enters a prism made of a solid transparent material. Suppose the prism's cross-section forms a perfect equilateral triangle with the base oriented horizontally, as shown in Fig. 11-13A. If the index of refraction equals 1.52000 for red light and 1.52500 for yellow light, what's the measure of the angle δ between rays of red and yellow light as they emerge from the prism, to the nearest hundredth of a degree? Assume that the surrounding medium constitutes a vacuum, so that its index of refraction equals exactly 1 (to as many significant figures as we need) for light of all colors.

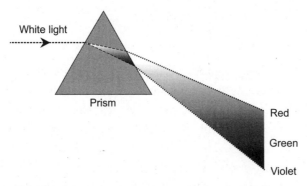

FIGURE 11-12 · Light rays of different colors refract at different angles through a prism.

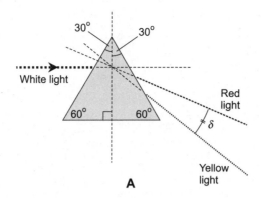

FIGURE 11-13A • Illustration for Problem 11-6.

![checkmark] SOLUTION _____

This problem requires quite a lot of calculation, so you'll need to use your powers of precision and patience to follow along! Let's solve it by going through the following three steps in order:

1. Trace the ray of red light all the way through the prism and determine the angle at which it exits the prism.
2. Trace the ray of yellow light in the same way.
3. Determine the difference in the two exit angles by subtracting one from the other.

   Refer to Fig. 11-13B. The ray of white light comes in horizontally, so the angle of incidence equals 30° at the prism surface. Let's consider this figure exact for purposes of calculation. (To minimize the risk of problems with cumulative rounding errors, let's carry out our intermediate calculations to six significant figures, rounding off at the end of the process.) We can calculate the measure of the angle ρ_1 that the ray of red light subtends relative to the normal line N_1, as the ray passes through the first surface into the prism, using the refraction formula. Step by step, we proceed as follows:

$$\sin \rho_1 / \sin 30.0000° = 1.00000 / 1.52000$$

$$\sin \rho_1 / 0.500000 = 0.657895$$

$$\sin \rho_1 = 0.500000 \times 0.657895$$

$$= 0.328948$$

$$\rho_1 = \text{Arcsin } 0.328948$$

$$= 19.2049°$$

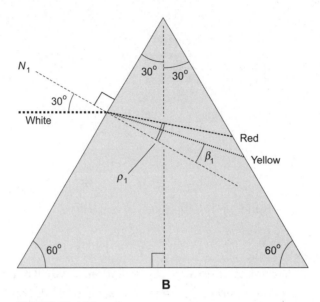

FIGURE 11-13B · Illustration for the first part of the solution to Problem 11-6.

Because the normal line N_1 to the first surface slants at 30.0000° relative to the horizontal, the ray of red light inside the prism slants at 30.0000° − 19.2049°, or 10.7951°, relative to the horizontal. Now look at Fig. 11-13C and compare it with Fig. 11-13B. The line normal to the second surface at the point where the red ray emerges (call it N_{2r}) slants at 30.0000° with respect to the horizontal just as line N_1 does, but upward instead of downward. Therefore, the angle of incidence ρ_2, at which the ray of red light strikes the inside second surface of the prism, equals 30.0000° + 10.7951°, or 40.7951°. Again we use the refraction formula, this time to find the angle ρ_3, relative to the normal N_{2r}, at which the ray of red light exits the second surface of the prism:

$$\sin \rho_3 / \sin 40.7951° = 1.52000 / 1.00000$$

$$\sin \rho_3 / 0.653356 = 1.52000$$

$$\sin \rho_3 = 0.653356 \times 1.52000$$
$$= 0.993101$$

$$\rho_3 = \text{Arcsin } 0.993101$$
$$= 83.2659°$$

Now we must repeat all of these calculations for the ray of yellow light. Refer again to Fig. 11-13B. The ray of white light comes in horizontally, so the angle of incidence equals 30.0000°, as before. We calculate the measure of angle β_1 that the ray of yellow light subtends relative to N_1, as it passes through the first surface into the prism, as follows, step by step:

$$\sin \beta_1 / \sin 30.0000° = 1.00000 / 1.52500$$

$$\sin \beta_1 / 0.500000 = 0.655738$$

$$\sin \beta_1 = 0.500000 \times 0.655738$$
$$= 0.327869$$

$$\beta_1 = \text{Arcsin } 0.327869$$
$$= 19.1395°$$

Because the normal line N_1 to the first surface slants at 30.0000° relative to the horizontal, the ray of yellow light inside the prism slants at 30.0000° − 19.1395°, or 10.8605°, relative to the horizontal. Now, once again, compare Fig. 11-13C with Fig. 11-13B. The line normal to the second surface at the point where the

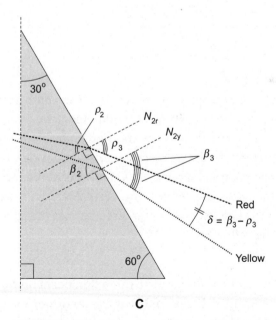

C

FIGURE 11-13C · Illustration for the second part of the solution to Problem 11-6.

yellow ray emerges (call it N_{2y}) slants at 30.0000° with respect to the horizontal, but in the opposite direction from line N_1, so the angle of incidence β_2, at which the ray of yellow light strikes the inside second surface of the prism, equals 30.0000° + 10.8605°, or 40.8605°. Again we use the refraction formula, this time to find the angle β_3, relative to the normal N_{2y}, at which the ray of yellow light exits the second surface of the prism:

$$\sin \beta_3 / \sin 40.8605° = 1.52500 / 1.00000$$

$$\sin \beta_3 / 0.654220 = 1.52500$$

$$\sin \beta_3 = 0.654220 \times 1.52500$$
$$= 0.997686$$

$$\beta_3 = \text{Arcsin } 0.997686$$
$$= 86.1015°$$

The difference $\beta_3 - \rho_3$ is the angle δ we seek, the angle between the rays of yellow and red light as they emerge from the prism. The measure of this angle, rounded off to the nearest hundredth of a degree, is

$$\beta_3 - \rho_3 = 86.1015° - 83.2659°$$
$$= 2.84°$$

PROBLEM 11-7

Suppose that we want to project a "rainbow" spectrum onto a screen, so that the spectrum image measures precisely 10 centimeters (cm) from the red band to the yellow band using the prism as configured in Problem 11-6. At what distance d from the screen should we place the prism? Consider the position of the prism as the intersection point between extensions of the red and yellow rays emerging from the prism, as shown in Fig. 11-14. Measure the distance d along the yellow ray to the nearest centimeter. Assume that the screen is oriented at a right angle to the yellow ray.

SOLUTION

We know that $\delta = 2.84°$ (accurate to three significant figures) from the solution to Problem 11-6. We're told that the length of the spectrum image from the red band to the yellow band, as it appears on the screen, equals exactly 10 cm. Let's go to some extra significant figures during the calcula-

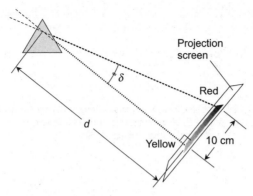

FIGURE 11-14 • Illustration for Problem 11-7.

tion process, rounding off at the end. We solve for *d* using right-triangle trigonometry as follows:

$$\tan 2.84° = 10.0000/d$$

$$0.049608 = 10.0000/d$$

$$0.049608 / 10.0000 = 1.00000/d$$

$$d = 10.0000 / 0.049608$$

$$= 202$$

We should place the screen at a distance of 202 cm from the prism, as we would measure it along the yellow ray of light according to the geometry of Fig. 11-14.

QUIZ

Refer to the text in this chapter if necessary. A good score is eight correct. Answers are in the back of the book.

1. Refer back to Fig. 11-6. Suppose the concave mirror has a focal length of 1.50 m. You stand at a distance of 8.35 m from the center of the mirror. How far from the center of the mirror does your real image lie?

    A. 1.33 m
    B. 1.67 m
    C. 1.83 m
    D. 5.57 m

2. Refer back to Fig. 11-9. Suppose that the upper medium has a refractive index of 1.05, while the lower medium has a refractive index of 1.75. Further suppose that ray R strikes the boundary between the media at an angle of incidence of 60° relative to the normal. At what angle, relative to the normal, does ray S emerge into the lower medium?

    A. 27°
    B. 31°
    C. 37°
    D. No such ray S exists because ray R undergoes total internal reflection at the boundary.

3. Refer again to Fig. 11-9. Suppose that we gradually increase the angle of incidence at which R strikes the boundary, relative to the normal, from the upper medium (starting at $\theta = 60°$ and getting larger until we approach 90°). What's the maximum possible angle, relative to the normal, at which ray S can emerge from the boundary into the lower medium?

    A. 27°
    B. 31°
    C. 37°
    D. No such ray S exists because ray R undergoes total internal reflection at the boundary when $\theta \geq 60°$.

4. Imagine a square room that measures 13.00 feet (13.00 ft) by 13.00 ft, with one mirrored wall, as shown in Fig. 11-15. You stand near wall W and hold a flashlight so that its bulb lies 2.750 ft away from W and 7.500 ft. away from the mirrored wall. You aim the flashlight horizontally at the mirrored wall, so that the center of its beam strikes the mirror at an angle of 67.50° relative to the mirror surface. The beam reflects from the mirror and encounters the wall opposite the mirror. What's the distance d?

    A. 11.24 ft
    B. 11.67 ft
    C. 12.00 ft

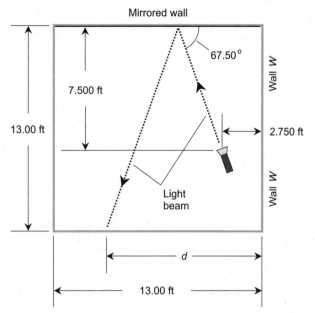

FIGURE 11-15 • Illustration for Quiz Question 4.

D. We can't answer this question as stated! The diagram, as shown here, doesn't represent the true state of affairs. The beam doesn't strike the wall opposite the mirror. It strikes the wall opposite *W*.

5. **When you derived the answer to Question 4, you had to rely on**

   A. Snell's law.
   B. the law of dispersion.
   C. the formula for the critical angle.
   D. the law of reflection.

6. **Based on the data in Table 11-1, what's the critical angle for a ray of light that passes from fresh liquid water into crown glass?**

   A. 41.2°
   B. 48.8°
   C. 61.0°
   D. No critical angle exists in this situation.

7. **Based on the data in Table 11-1, what's the critical angle for a ray of light that passes from crown glass into fresh liquid water?**

   A. 41.2°
   B. 48.8°
   C. 61.0°
   D. No critical angle exists in this situation.

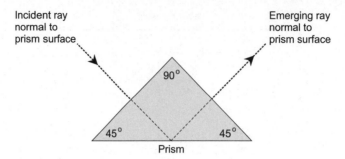

Incident ray normal to prism surface

Emerging ray normal to prism surface

90°

45°   45°

Prism

FIGURE 11-16 · Illustration for Quiz Question 8.

8. Some optical devices use prisms to reflect light at right angles. Figure 11-16 shows a cross section of such a prism. The incident and emerging rays strike the boundaries along normal lines so refraction doesn't occur. Total internal reflection takes place at the base of the prism, so that it acts as a high-quality mirror. Assuming that the surrounding medium is air with a refractive index of 1.00000, what's the minimum refractive index that the prism shown in Fig. 11-16 must have in order to ensure total internal reflection at its base, rounded off to four significant figures? Assume that the angles shown in the figure are mathematically exact.

A. 1.732
B. 1.414
C. 1.333
D. 1.250

9. Imagine that a ray of red light and a ray of violet light both strike the surface of a flat pane of clear, solid material at an incidence angle of 25.0000° with respect to the normal. The rays follow exactly the same path through a vacuum, striking the solid surface at the same point. The solid substance has a refractive index of 1.480 for red light and 1.520 for violet light. As the red and violet rays emerge inside the solid medium, by what angle do they diverge from each other, accurate to the nearest hundredth of a degree?

A. 0.33°
B. 0.45°
C. 0.88°
D. 1.36°

10. Suppose that you repeat the experiment described in Question 9, but instead of an incidence angle of 25.0000°, the rays strike the surface at 85.0000° with respect to the normal. As the red and violet rays emerge inside the medium, by what angle do they diverge from each other, accurate to the nearest hundredth of a degree?

A. 0.33°
B. 0.45°
C. 0.88°
D. 1.36°

chapter **12**

Trigonometry on a Sphere

All the trigonometry that we've done so far has taken place on flat surfaces, or in space where all the lines follow straight paths. But in the real world—in particular, on the surface of the earth—lines rarely run straight!

CHAPTER OBJECTIVES

In this chapter, you will

- Define great circles and arcs on the surface of a sphere.
- Scrutinize the concepts of latitude and longitude in greater detail than you did in Chap. 7.
- Learn how triangles and other polygons behave when confined to a spherical surface.
- Discover the laws of sines and cosines for spherical surfaces.
- Use spherical trigonometry to solve practical problems in navigation.

The Global Grid

When we confine our "geometric universe" to the surface of a sphere, no such thing as a straight line or line segment exists. The closest thing to a straight line in this environment is a *great circle* or *geodesic*. The closest thing to a straight line segment is an *arc of a great circle* or *geodesic arc*.

Great Circles

The surface of a sphere comprises all points in three-dimensional (3D) space that lie equally far from some defined center point P. All paths on the surface of a sphere form curves. If Q and R represent any two points on the surface of a sphere, then the straight line segment QR inevitably cuts through the interior of the sphere, even if the points lie close to each other.

We can express the shortest distance between two points Q and R on the surface of a sphere (rather than through its interior) along a geodesic arc, which forms part of a circle on the sphere that has P at its center (Fig. 12-1). This rule never fails for a "perfectly round ball." The earth, averaged out over all of its terrain, comes close enough to a "perfectly round ball" to let us apply this rule to our planet.

TIP   *Henceforth, when we say "the surface of the earth," we mean "the sphere corresponding to the surface of the earth at sea level." Let's ignore all the local irregularities such as hills, valleys, mountains, or buildings.*

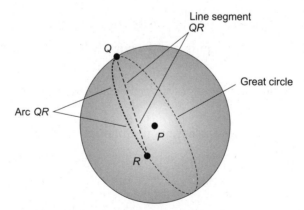

FIGURE 12-1 · The shortest path over the surface of a sphere between points Q and R follows an arc centered at P, the center of the sphere. The length of arc QR always exceeds the length of the straight line segment QR.

Latitude Revisited

We can define latitude as an angle, going either toward the north (positive) or toward the south (negative), with respect to any point on a great circle representing the *equator*, the set of surface points equidistant from the north geographic pole and the south geographic pole. At the geographic poles, the earth's rotational axis intersects the surface. The latitude, commonly denoted θ, can have values as large as $+90°$ or as small as $-90°$, inclusive. In formal symbolic terms, we write this fact as

$$-90° \leq \theta \leq +90°$$

We can also write

$$90° \text{ S} \leq \theta \leq 90° \text{ N}$$

where "S" stands for "south" and "N" stands for "north."

TIP   *For any point on the earth's surface, its latitude equals the angle between the plane containing the equator and a ray whose back-end (originating) point lies at the earth's center.*

Longitude Revisited

We can define longitude as an angle, going either toward the east (positive) or toward the west (negative), with respect to a great circle called the *prime meridian*. We must always express or measure longitude angles going around the equator, or around any circle on the surface of the earth's surface parallel to the equator. The prime meridian has its end points at the north pole and the south pole, and it intersects the equator at a right angle.

Several generations ago, geographers and map-makers decided that the town of Greenwich, England, would receive the distinction of having the prime meridian pass through it. For that reason, some people call the prime meridian the *Greenwich meridian*. Angles of longitude, denoted ϕ, can range between $-180°$ and $+180°$, not including the negative value. In symbolic terms, we can write

$$-180° < \phi \leq +180°$$

We can also write

$$180° \text{ W} < \phi \leq 180° \text{ E}$$

where "W" stands for "west" (negative longitude values) and "E" stands for "east" (positive longitude values).

> **TIP** *For any point on the earth's surface, its longitude equals the angle between the plane containing the prime meridian and the plane containing the great circle that runs through both poles and also through the point in question.*

> **TIP** *According to popular legend, the people of France expressed keen disappointment when "the experts" decided to run the prime meridian through England. The French apparently wanted that exalted geodesic arc to pass through Paris! If the French had gotten their way, you'd be reading about the "Paris Meridian" right now.*

Parallels

For any given angle θ larger than $-90°$ and smaller than $+90°$, we can find infinitely many points on the earth's surface with latitude θ. This set of points always forms a circle parallel to the equator, so we call it a *parallel* (Fig. 12-2A). The exceptions occur at the extremes $\theta = -90°$ and $\theta = +90°$. These latitude angles correspond to single points at the south geographic pole and the north geographic pole, respectively.

The radius of a given parallel depends on its latitude. When $\theta = 0°$, the parallel corresponds to the equator, and its radius equals the earth's radius. If we imagine the earth as a perfect sphere, then its radius equals approximately 6371 kilometers. That's the radius of the parallel corresponding to $\theta = 0°$. For other values of θ, we can calculate the radius r (in kilometers) of the parallel as

$$r = 6371 \cos \theta$$

The earth's circumference equals approximately $6371 \times 2\pi$, or 4.003×10^4 kilometers, so we can calculate the circumference k (in kilometers) of the parallel at latitude θ using the formula

$$k = (4.003 \times 10^4) \cos \theta$$

Meridians

For any given angle ϕ such that $-180° < \phi \leq +180°$, there exists a set of points on the earth's surface such that all the points have longitude ϕ. This set of points forms a half-circle (not including either of the end points) whose center coincides with the center of the earth, and that intersects the equator at a right angle, as shown in Fig. 12-2B. We call every such open half-circle a *meridian*. The end points of any meridian (which technically aren't part of the meridian itself) correspond to the south geographic pole and the north geographic pole. The poles themselves have undefined longitudes.

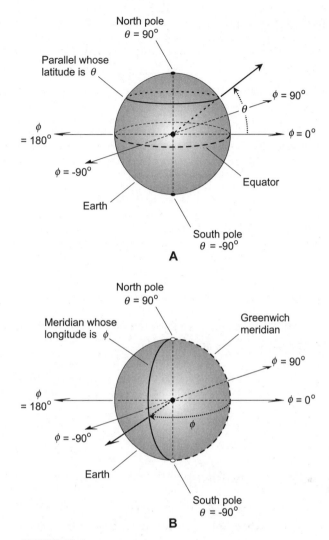

FIGURE 12-2 · At A, parallels form circles representing points of equal latitude. At B, meridians form half-circles representing points of equal longitude.

All meridians have the same radius, which equals the radius of the earth, approximately 6371 kilometers. All meridians converge at the poles. The distance between any two fixed meridians, as measured along a particular parallel, depends on the latitude of that parallel. The distance between equal-latitude points on any two meridians ϕ_1 and ϕ_2 is greatest at the equator, decreases as the latitude increases negatively or positively, and approaches zero as the latitude approaches $-90°$ (the south pole) or $+90°$ (the north pole).

Distance per Unit of Latitude

As measured over the earth's surface along any fixed meridian (in a north-south direction), the distance $d_{\text{lat-deg}}$ per degree of latitude remains constant. We can calculate it by dividing the circumference of the earth by 360. Knowing that the earth's radius equals 6371 kilometers, we calculate its circumference as $2\pi \times 6371$, or 4.003×10^4 kilometers. If we express $d_{\text{lat-deg}}$ in kilometers per degree (km/deg), then

$$d_{\text{lat-deg}} = (4.003 \times 10^4)/360$$
$$= 111.2 \text{ km/deg}$$

The distance $d_{\text{lat-min}}$ in kilometers per minute of arc (km/arc-min) can be obtained by dividing the above figure by exactly 60. We get

$$d_{\text{lat-min}} = 111.2/60.00$$
$$= 1.853 \text{ km/arc-min}$$

We can calculate the distance $d_{\text{lat-sec}}$ in kilometers per second of arc (km/arc-sec) by dividing by exactly 60 once again, getting

$$d_{\text{lat-sec}} = 1.853/60.00$$
$$= 0.03088 \text{ km/arc-sec}$$

We might do better to state this last figure as $d_{\text{lat-sec}} = 30.88$ meters per second of arc (m/arc-sec) of latitude.

Distance per Unit of Longitude

As measured along the equator, the distances $d_{\text{lon-deg}}$ (distance per degree of longitude), $d_{\text{lon-min}}$ (distance per arc minute of longitude), and $d_{\text{lon-sec}}$ (distance per arc second of longitude), in kilometers, can be found according to the same formulas as those for the distance per unit latitude. Once again, the formulas are

$$d_{\text{lon-deg}} = (4.003 \times 10^4)/360$$
$$= 111.2 \text{ km/deg}$$

$$d_{\text{lon-min}} = 111.2/60.00$$
$$= 1.853 \text{ km/arc-min}$$

$$d_{\text{lon-sec}} = 1.853/60.00$$
$$= 0.03088 \text{ km/arc-sec}$$
$$= 30.88 \text{ m/arc-sec}$$

These formulas don't work when we want to determine the east-west distance between any two particular meridians along a parallel other than the equator. In order to calculate those distances, we must multiply the above values by the cosine of the latitude θ at which we make the measurement. When we modify the formulas that way, we get

$$d_{\text{lon-deg}} = 111.2 \cos \theta$$

for kilometers per degree of longitude, or

$$d_{\text{lon-min}} = 1.853 \cos \theta$$

for kilometers per arc minute of longitude, or

$$d_{\text{lon-sec}} = 0.03088 \cos \theta$$

for kilometers per arc second of longitude. We can write the last formula to express the distance in meters per arc second as

$$d_{\text{lon-sec}} = 30.88 \cos \theta$$

PROBLEM 12-1

Imagine a certain large warehouse, with a square floor measuring 100 meters on a side, located in a community at 60°0'0" north latitude. Suppose that the warehouse is oriented "kitty-corner" to the points of the compass, so that its sides run northeast-by-southwest and northwest-by-southeast. What's the difference in longitude, expressed in seconds of arc, between the west corner and the east corner of the warehouse? (Because the longitude difference angle is small, let's ignore the tiny theoretical difference between the straight-line distance and the arc length over the sphere representing the earth.)

SOLUTION

Figure 12-3 illustrates this situation. Let $\Delta\phi_{\text{wh}}$ represent the difference in longitude between the west and east corners of the warehouse. (The Δ symbol in this context is the uppercase Greek letter delta, which means "the difference in." It doesn't stand for a triangle as it does in geometry.)

Let's find the distance in meters between corners of the warehouse. It's equal to $100 \times 2^{1/2}$, or approximately 141.4, meters. Now let's find out how many meters exist per arc second at 60°0'0" north latitude. We calculate

$$d_{\text{lon-sec}} = 30.88 \cos 60°0'0"$$
$$= 30.88 \times 0.5000$$
$$= 15.44 \text{ m/arc-sec}$$

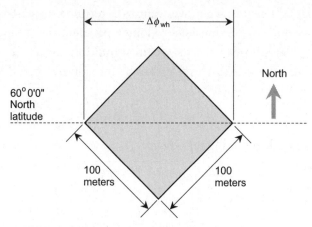

FIGURE 12-3 · Illustration for Problems 12-1 and 12-2.

In order to calculate $\Delta\phi_{wh}$, the number of arc seconds of longitude between the east and west corners of the warehouse, we must divide 141.4 meters by 15.44 meters per arc second, getting

$$\Delta\phi_{wh} = 141.4 / 15.44$$
$$= 9.16''$$

We should round our answer off to three significant figures because that's the extent of the accuracy of our input data (100 meters along each edge of the warehouse). If we want to express this longitude difference in degrees, minutes, and seconds, we write

$$\Delta\phi_{wh} = 0°0'9.16''$$

 **PROBLEM 12-2**

What's the difference in latitude, expressed in seconds of arc, between the north and the south corners of the warehouse described above?

✔ SOLUTION

We already know that the straight-line distance between corners of the warehouse equals 141.4 meters. We also know that each arc second contains 30.88 meters of distance, as measured in a north-south direction, at any latitude. Let $\Delta\theta_{wh}$ represent the difference in latitude between the north and the south corners. We divide 141.4 meters by 30.88 meters per arc second, obtaining

$$\Delta\theta_{wh} = 141.4/30.88$$
$$= 4.58''$$

Again, we round off to three significant figures because that's the extent of the accuracy of our input data (100 meters along each edge of the warehouse). If we want to express this longitude difference in degrees, minutes, and seconds, we write

$$\Delta\theta_{wh} = 0°0'4.58''$$

Arcs and Triangles

We've defined latitude and longitude on the surface of the earth, and we know how to find the differences in latitude and longitude between points along north-south and east-west paths. Now let's learn how to find the distance between any two points on the earth's surface, as measured along a great circle passing through them both.

Which Arc?

Two different great-circle arcs can connect any two points on a sphere. One arc goes halfway around the sphere or farther, and the other arc goes halfway around the sphere or less. When we put these two arcs together, we get a complete great circle that passes through both points. The shorter of the two arcs represents the most direct possible route, over the surface of the earth, between the two points. When we talk about the "distance between two points on a sphere," we mean to say the distance as measured along the shorter of the two great-circle arcs (also called the *minor geodesic*) connecting them.

TIP *If two points lie at* antipodes *on the sphere—exactly opposite each other— then we can take any great circle arc between them that we want in order to define the over-the-surface distance between the points. All geodesic arcs connecting a sphere's antipodes have the same length, equal to half the sphere's circumference.*

Still Struggling

Consider a practical example of the above-mentioned principle. You can get from New York to Los Angeles more easily by flying west across North America than by flying east over the Atlantic Ocean, Africa, the Indian Ocean, Australia, and the Pacific Ocean. Any reasonable person would define the distance between those two cities along the shorter or minor geodesic, not along the longer one (the *major geodesic*)!

Spherical Triangles

We can construct a *spherical triangle* on the basis of three *vertex* points on the surface of a sphere. Imagine three such points Q, R, and S. Let P represent the center of the sphere. The spherical triangle, denoted $\Delta_{\text{sph}} QRS$, has sides q, r, and s opposite the vertex points Q, R, and S respectively, as shown in Fig. 12-4. (Here, the uppercase Greek delta stands for the word "triangle," not "the difference in" as it did earlier in this chapter.)

Each side of the spherical triangle constitutes a geodesic arc spanning less than 360°, so each side must go less than once around the sphere. We can't

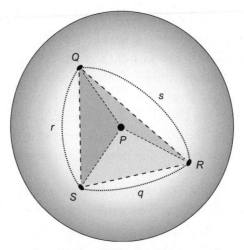

FIGURE 12-4 · Each side of a spherical triangle forms an arc of a great circle.

"allow" a spherical triangle to have any side that goes all the way around the sphere, or farther.

For any spherical triangle, we can identify three ordinary plane triangles defined by the vertices of the spherical triangle and the center of the sphere. In Fig. 12-4, these ordinary triangles are ΔPQR, ΔPQS, and ΔPRS. All three of these triangles define planes in the 3D space within the sphere; we can call them plane PQR, plane PQS, and plane PRS. As things work out, the following things hold true:

- Plane PQR contains arc s
- Plane PQS contains arc r
- Plane PRS contains arc q

Spherical Polygons

We can define a *spherical polygon* on the basis of three or more points that all lie on the surface of a sphere. Each side of a spherical polygon forms a geodesic arc spanning less than 360°, so each side must go less than once around the sphere. We can't "allow" a spherical polygon with any side that goes all the way around the sphere, or farther.

Spherical Angles

The sides of any spherical triangle form curves, not straight lines. We call the interior angles of a spherical triangle *spherical angles*. Let's symbolize the words "spherical angle" as $\angle_{sph}$. Imagine the flat interiors of the three triangles ΔPQR, ΔPQS, and ΔPRS, as shown in Fig. 12-4. We can define the three spherical angles between the arcs as follows:

1. The angle between the flat interiors of ΔPQR and ΔPQS, which intersect in line segment PQ, has the same measure as the spherical angle between arcs r and s, whose vertex lies at point Q.

2. The angle between the flat interiors of ΔPQR and ΔPRS, which intersect in line segment PR, has the same measure as the spherical angle between arcs q and s, whose vertex lies at point R.

3. The angle between the flat interiors of ΔPQS and ΔPRS, which intersect in line segment PS, has the same measure as the spherical angle between arcs q and r, whose vertex lies at point S.

Angular Sides

The sides q, r, and s of the spherical triangle of Fig. 12-4 are often defined in terms of their *arc angles* ($\angle SPR$, $\angle QPS$, and $\angle RPQ$, respectively), rather than in terms of their actual lengths in linear units.

If we know the radius of a sphere (call it r_{sph}), then the length of an arc on the sphere, in the same linear units as we use to measure the radius of the sphere, equals the angular measure of the arc (in radians) multiplied by r_{sph}. Suppose that we let the symbols $|q|$, $|r|$, and $|s|$ represent the lengths of the arcs q, r, and s of $\Delta_{sph}QRS$ in linear units, while we denote their extents in angular radians simply as q, r, and s. Then we can relate the arc lengths and the angles with three formulas, as follows:

$$|q| = r_{sph}\, q$$

$$|r| = r_{sph}\, r$$

$$|s| = r_{sph}\, s$$

On the surface of the earth, we can express the linear lengths (in kilometers) of the sides of the spherical triangle $\Delta_{sph}QRS$ as follows:

$$|q| = 6371\, q$$

$$|r| = 6371\, r$$

$$|s| = 6371\, s$$

PROBLEM 12-3

A great-circle arc on the earth's surface has a measure of 1.500 rad. What's the length of this arc in kilometers?

SOLUTION

We multiply 6371, the radius of the earth in kilometers, by 1.500, obtaining 9556.5 kilometers. We should round this result off to 9557 kilometers because the input data only goes to four significant figures of accuracy.

PROBLEM 12-4

Describe and draw an example of a spherical triangle on the earth's surface in which two of the interior spherical angles are right angles. Then describe and draw an example of a spherical triangle on the surface of the earth in which all three interior spherical angles are right angles.

☑ SOLUTION

To solve the first part of the problem, consider the spherical triangle $\Delta_{sph}QRS$ such that points Q and R lie on the equator, and point S lies at the north geographic pole (Fig. 12-5A). The two interior spherical angles $\angle_{sph}SQR$ and $\angle_{sph}SRQ$ form right angles because sides SQ and SR of $\Delta_{sph}QRS$ lie along meridians, while side QR lies along the equator. (All meridian arcs intersect the equator at right angles.)

To solve the second part of the problem, we construct $\Delta_{sph}QRS$ such that points Q and R lie along the equator, separated by 90° of longitude

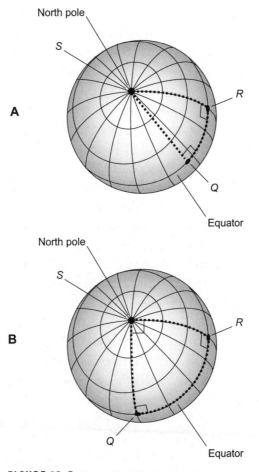

FIGURE 12-5 · Illustrations for Problem 12-4.

(Fig. 12-5B). In this scenario, the measure of $\angle_{sph}QSR$, whose vertex lies at the north pole, equals 90°. We already know that the measures of angles $\angle_{sph}SQR$ and $\angle_{sph}SRQ$ both equal 90°. Therefore, we can conclude that all three of the interior spherical angles of $\Delta_{sph}QRS$ are right angles.

The Expanding Triangle

Imagine what happens to an *equilateral spherical triangle* that starts out tiny and grows larger. An equilateral spherical triangle has geodesic-arc sides that lie on the sphere's surface and cover equal angular measure (called *angular extent*) with respect to the sphere's center, and interior spherical angles of equal measure. An equilateral spherical triangle on the earth that measures 1 arc second on a side is almost exactly the same as a plane equilateral triangle whose sides are 30.88 meters long. The sum of the interior spherical angles, if we measure them with a surveyor's apparatus, appears to be 180°, and each interior spherical angle appears as an ordinary angle that measures 60°. We can calculate the interior area and the perimeter using the formulas for triangles in a flat plane.

As the equilateral spherical triangle grows, the measure of each interior spherical angle increases. When each side has a length equivalent to 1/4 of a great circle (the angular extent of each side equals $\pi/2$ rad), then each interior spherical angle measures 90°, and the sum of the measures of the interior spherical angles equals three times 90°, or 270°. Figure 12-5B shows an example. As the spherical triangle keeps on growing, it eventually attains a perimeter equal to the circumference of the earth. Then each side has an angular extent of $2\pi/3$ rad. The spherical triangle has, in fact, evolved into a great circle! Its interior area has grown to half the surface area of the planet—in which case you and I might disagree about which half of the earth's surface lies inside the triangle and which half lies outside! The formulas for the perimeter and interior area of a plane triangle don't work for a spherical triangle that covers a substantial part of the globe.

Now think about what happens if the equilateral spherical triangle continues to "grow" beyond the size at which it girdles the earth. The lengths of the sides get shorter, not longer, even though the measures of the interior spherical angles, and the interior area of the spherical triangle, keep increasing. Ultimately, our equilateral spherical triangle becomes so "large" that the three vertices lie close together again, perhaps only 1 arc second apart. We have something that looks like the tiny triangle we started out with. But wait! The vertices might be the same, but the triangle is vastly different! The perimeter of the "huge new triangle" equals the perimeter of the original tiny one, but the interior area of the "huge new triangle" is almost as great as the surface area of the whole sphere. The inside of this "huge new triangle" looks like the outside of the "tiny old triangle," and the outside of the new triangle looks like the inside of the old one. The interior spherical angles don't appear to be 60°, as they did in the "tiny old triangle," but instead appear to measure 300°. We must express or measure them "the long way around the sphere." The sum of their measures appears to be 900°, not 180°!

The Long Way Around

Imagine a spherical triangle whose vertices lie near each other, but whose sides go the long way around the globe, as shown in Fig. 12-6. For a few moments, let's liberate ourselves of the constraint that each side of a spherical polygon must make less than one circumnavigation of the sphere. In that case, any spherical polygon can have sides that go more than once around, maybe hundreds of times, maybe millions of times. Can you "see in your mind's eye" what constitutes the interior of an "equilateral spherical supertriangle" whose sides each make, say, 72.9 trips around the surface of its "parent sphere"?

TIP *You might think of the above-described "equilateral spherical supertriangle" as a globe wrapped up in layer upon layer of infinitely thin tape. And what about the exterior? Perhaps you can think of the globe wrapped up in infinitely thin tape—made out of antimatter!*

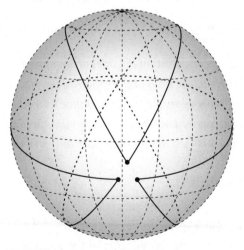

FIGURE 12-6 • A spherical triangle in which each side goes nearly all the way around the globe.

Still Struggling

We can have fun playing "mind games" such as the foregoing, but they get confusing when we try to make them logical. From now on, whenever we talk about a spherical polygon, let's limit its size as follows:

- The perimeter can't exceed the circumference of the sphere.
- The interior area can't exceed half the surface area of the sphere.

Let's tag any object that violates either of these rules as "illegal" or "nonstandard," unless we clearly state our intention to allow the exception.

Spherical Law of Sines

For any spherical triangle, a relationship called the *spherical law of sines* exists among the angular extents (in radians) of the sides and the measures of the interior spherical angles. Let $\Delta_{sph}QRS$ represent a spherical triangle whose vertices lie at points Q, R, and S. Let's call the angular extents of the sides opposite each vertex point, expressed in radians, by the names q, r, and s respectively, as shown in Fig. 12-7. Let's assign the interior spherical angles $\angle_{sph}RQS$, $\angle_{sph}SRQ$, and $\angle_{sph}QSR$ the labels ψ_q, ψ_r, and ψ_s respectively. (The symbol ψ is an italicized, lowercase Greek letter psi; we use it instead of θ to indicate spherical angles.) In this situation, we can state the spherical law of sines as

$$(\sin q)/(\sin \psi_q) = (\sin r)/(\sin \psi_r) = (\sin s)/(\sin \psi_s)$$

TIP   *The sines of the angular extents of the sides of any spherical triangle maintain a constant ratio relative to the sines of the spherical angles opposite those sides. The spherical law of sines bears some resemblance to the law of sines for ordinary triangles in a flat plane. (Look back at the original law of sines, described in Chap. 9 and illustrated in Fig. 9-10, and compare that one with this one.)*

Spherical Law of Cosines

The *spherical law of cosines* provides another useful rule for dealing with spherical triangles. Let's define a spherical triangle as we did above and in Fig. 12-7. Suppose that we know the angular extents of two of the sides, say q and r, and

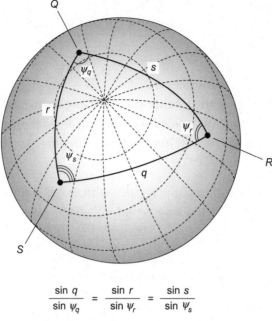

$$\frac{\sin q}{\sin \psi_q} = \frac{\sin r}{\sin \psi_r} = \frac{\sin s}{\sin \psi_s}$$

$$\cos s = \cos q \ \cos r + \sin q \ \sin r \ \cos \psi_s$$

FIGURE 12-7 · The law of sines and the law of cosines for spherical triangles.

the measure of the spherical angle ψ_s between them. In that case, we can calculate the cosine of the angular extent of the third side s with the formula

$$\cos s = \cos q \cos r + \sin q \sin r \cos \psi_s$$

TIP *The spherical law of cosines doesn't look much like the law of cosines for triangles in a flat plane, so you can forget about trying to draw a meaningful comparison!*

Equilateral Spherical Triangles

In ordinary *Euclidean geometry* (geometry in a flat plane), if a triangle has sides all equally long, then the interior angles all have the same measure. The converse of this statement also holds true. If the interior angles of a triangle all have equal measure, then the sides all have the same length.

An analogous principle applies to equilateral triangles on a sphere. If a spherical triangle has sides all of the same angular extent, then the interior spherical angles all have equal measure. Conversely, if the interior spherical angles of all have equal measure, then the sides all have the same angular extent.

**PROBLEM 12-5**

Calculate the measures of the interior spherical angles, in degrees, of an equilateral spherical triangle whose sides each have an angular span of 1.00000 rad. Express the results to the nearest hundredth of a degree.

SOLUTION

Let's call the spherical triangle $\triangle_{sph}QRS$, with vertex points called Q, R, and S, and sides called q, r, and s, such that

$$q = r = s = 1.00000 \text{ rad}$$

Then we can calculate the following cosines and sines:

$$\cos q = 0.540302$$

$$\cos r = 0.540302$$

$$\cos s = 0.540302$$

$$\sin q = 0.841471$$

$$\sin r = 0.841471$$

Now let's plug these values into the formula for the spherical law of cosines to solve for $\cos \psi_s$, where ψ_s represents the measure of the angle opposite side S. The general formula, once again, is

$$\cos s = \cos q \cos r + \sin q \sin r \cos \psi_s$$

Inputting the known values gives us

$$0.540302 = 0.540302 \times 0.540302 + 0.841471 \times 0.841471 \times \cos \psi_s$$

which solves to

$$\cos \psi_s = (0.540302 - 0.291926)/0.708073$$

$$= 0.350777$$

Therefore, we know that

$$\psi_s = \text{Arccos } 0.350777$$

$$= 69.4652°$$

Rounding to the nearest hundredth of a degree yields a spherical angle of 69.47°. Because the triangle is equilateral, we know that all three interior spherical angles ψ have the same measure: approximately 69.47°.

PROBLEM 12-6

Consider an equilateral spherical triangle $\triangle_{sph}QRS$ on the surface of the earth, whose sides each measure 1.00000 rad in angular extent, as in the previous problem. Suppose that vertex Q lies at the north pole (latitude +90.0000°) and vertex R lies on the Greenwich meridian (longitude 0.0000°). Suppose that vertex S lies somewhere in the western hemisphere, so it has a negative longitude value. What are the latitude and longitude coordinates of each vertex to the nearest hundredth of a degree?

✔ SOLUTION

Figure 12-8 shows this situation. We know that the latitude of point Q (Lat Q) equals +90.0000°. We can't define the longitude of Q (Lon Q) because Q lies precisely where all the longitude arcs converge. We know that Lon $R = 0.0000°$. We don't yet know Lat R, but we can say that it must

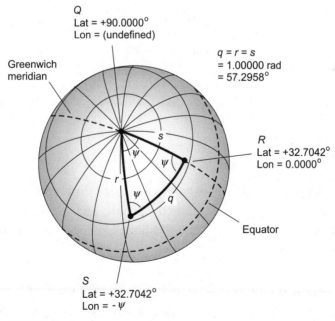

FIGURE 12-8 · Illustration for Problem 12-6.

be $+90.0000°$ minus the angular extent of side s. This value equals $+90.0000° - 1.00000$ rad. When we note that 1.00000 rad $= 57.2958°$ (accurate to six significant figures), we can calculate

$$\text{Lat } R = +90.0000° - 57.2958°$$

$$= +32.7042°$$

Rounded off to the nearest hundredth of a degree, Lat $R = +32.70°$. This value must also represent the latitude of vertex S because the angular extent of side r equals the angular extent of side s. (Remember, we're dealing with an equilateral spherical triangle!) The longitude of vertex S equals the negative of the measure of the interior spherical angle at the pole, or $-\psi$. We can infer from the solution to Problem 12-5 that $\psi = 69.47°$. Therefore, we have the following coordinates for the vertices of $\triangle_{\text{sph}} QRS$, rounded off to the nearest hundredth of a degree:

$$\text{Lat } Q = +90.00°$$

$$\text{Lon } Q = \text{(undefined)}$$

$$\text{Lat } R = +32.70°$$

$$\text{Lon } R = 0.00°$$

$$\text{Lat } S = +32.70°$$

$$\text{Lon } S = -69.47°$$

Global Navigation

Spherical trigonometry, when done on the surface of the earth, serves mariners, aviators, aerospace engineers, and military people, and provides the basis for determining great-circle distances and headings. Let's work out four problems relating to global navigation.

TIP *You can solve the following problems quickly with the help of a good computer application designed for the purpose, but you can get more familiar with* *the principles of global trigonometry by doing the calculations manually. So get ready to "do things the long way"!*

PROBLEM 12-7

Consider two points R and S on the earth's surface. Suppose that the points have the following coordinates:

$$\text{Lat } R = +50.00°$$

$$\text{Lon } R = +42.00°$$

$$\text{Lat } S = -12.00°$$

$$\text{Lon } S = -67.50°$$

What's the angular distance between points R and S, expressed to the nearest hundredth of a radian?

SOLUTION

Let's construct a spherical triangle with R and S at two of the vertices, and the third vertex at the geographic north pole. Let's call that vertex point Q. (The south pole will work too, but it's more awkward because the sides of the spherical triangle extend over greater portions of the globe.) We label the sides opposite each vertex q, r, and s. Therefore, we seek the angular distance q, as shown in Fig. 12-9. We can use the spherical law of cosines to determine q, provided that we can figure out three things:

1. The measure of ψ_q, the spherical angle at vertex Q
2. The angular extent of side r
3. The angular extent of side s

The measure of ψ_q equals the absolute value of the difference in the longitudes of the two points R and S. Doing the arithmetic, we get

$$\psi_q = |\text{Lon } R - \text{Lon } S|$$
$$= |+42.00° - (-67.50°)|$$
$$= 42.00° + 67.50°$$
$$= 109.50°$$

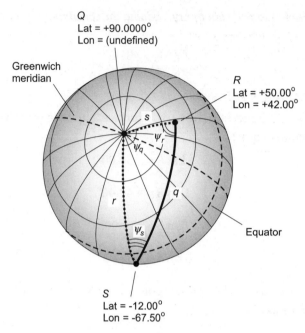

FIGURE 12-9 · Illustration for Problems 12-7 through 12-10.

The angular extent of side *r* equals the absolute value of the difference in the latitudes of points *Q* and *S*. The arithmetic process in this case yields

$$r = |\,\text{Lat } Q - \text{Lat } S\,|$$
$$= |+90.00° - (-12.00°)|$$
$$= 90.00° + 12.00°$$
$$= 102.00°$$

The angular extent of side *s* equals the absolute value of the difference in the latitudes of points *Q* and *R*. Going through some more arithmetic, we obtain

$$s = |\,\text{Lat } Q - \text{Lat } R\,|$$
$$= |+90.00° - (+50.00°)|$$
$$= 90.00° - 50.00°$$
$$= 40.00°$$

Now that we know the values of *r*, *s*, and ψ_q, the spherical law of cosines tells us that

$$\cos q = \cos r \cos s + \sin r \sin s \cos \psi_q$$

When we solve for q and round it off, we end up with

$q =$ Arccos (cos r cos s + sin r sin s cos ψ_q)

$\quad=$ Arccos (cos 102.00° × cos 40.00° + sin 102.00° × sin 40.00° × cos 109.50°)

$\quad=$ Arccos [(−0.20791) × 0.76604 + 0.97815 × 0.64279 × (−0.33381)]

$\quad=$ Arccos (−0.36915)

$\quad=$ 1.95 rad

 PROBLEM 12-8

Imagine that, as an aircraft pilot, you want to fly a great circle route from point R to point S in the scenario of Problem 12-7 and Fig. 12-9. What's the distance, in kilometers, that your aircraft will have to fly? You can neglect the small discrepancy caused by the fact that the aircraft travels several kilometers above the earth's surface, a small fraction of the earth's actual radius. Express the answer to three significant figures.

SOLUTION

You can multiply the angular distance, 1.95 rad, by 6371 kilometers, getting a distance of 1.24×10^4 kilometers, accurate to three significant figures.

 PROBLEM 12-9

Suppose, once again, that you want to fly your aircraft over a great circle route from point R to point S in the scenario of Problem 12-7 and Fig. 12-9. What initial azimuth heading should you use, once the aircraft approaches cruising altitude after taking off from point R? Express the answer to the nearest degree.

 SOLUTION

In order to determine this value, you must figure out the measure of angle ψ_r in degrees. The initial azimuth heading equals 360° − ψ_r because side s of $\triangle_{sph}QRS$ runs directly northward, or toward azimuth 0°, from point R. The spherical law of sines tells you the following fact concerning $\triangle_{sph}QRS$:

$$(\sin q)/(\sin \psi_q) = (\sin r)/(\sin \psi_r)$$

You can solve this equation for ψ_r by manipulating the above expression and then finding the Arcsine. The general formula is

$$\sin \psi_r = [(\sin r)(\sin \psi_q)]/(\sin q)$$

which you can convert to

$$\psi_r = \text{Arcsin}\{[(\sin r)(\sin \psi_q)]/(\sin q)\}$$

You already have the following information from the solution to Problem 12-7:

$$q = 1.95 \text{ rad}$$

$$\psi_q = 109.5°$$

$$r = 102.00°$$

When you input these numbers to the previous formula, you get

$$\psi_r = \text{Arcsin}\{[(\sin 102.00°)(\sin 109.5°)]/(\sin 1.95 \text{ rad})\}$$

$$= \text{Arcsin}[(0.97815 \times 0.94264)/0.92896]$$

$$= \text{Arcsin}\ 0.99255$$

$$= 83.00°$$

Your initial azimuth heading, following a great circle from point R to point S, should equal $360° - 83.00°$, or $277°$, expressed to the nearest degree. That's $7°$ north of west.

PROBLEM 12-10

Suppose that you want to fly an aircraft along a great circle route from point S to point R in the scenario of Problem 12-7 and Fig. 12-9 (exactly the return route from that described in Problem 12-9). What initial azimuth heading should you use, once the aircraft approaches cruising altitude after taking off from point S? Express the answer to the nearest degree.

✔ SOLUTION

You must figure out the measure of angle ψ_s in degrees. The initial azimuth heading equals ψ_s because side r of $\Delta_{sph}QRS$ runs directly northward, or toward azimuth 0°, from point S. According to the spherical law of sines, you know that

$$(\sin q)/(\sin \psi_q) = (\sin s)/(\sin \psi_s)$$

You can solve this equation for ψ_s by manipulating the above expression and then finding the Arcsine. The general formula is

$$\sin \psi_s = [(\sin s)(\sin \psi_q)]/(\sin q)$$

which converts directly to

$$\psi_s = \text{Arcsin}\{[(\sin s)(\sin \psi_q)]/(\sin q)\}$$

You already know the following information, having solved Problem 12-7:

$$q = 1.95 \text{ rad}$$

$$\psi_q = 109.5°$$

$$s = 40.00°$$

Plugging in the numbers gives you

$$\psi_s = \text{Arcsin}\{[(\sin 40.00°)(\sin 109.5°)]/(\sin 1.95 \text{ rad})\}$$
$$= \text{Arcsin}[(0.64279 \times 0.94264)/0.92896]$$
$$= \text{Arcsin } 0.65226$$
$$= 40.71°$$

Your initial heading, following a great circle from point S to point R, should be 41° to the nearest degree. That's 41° east of north.

QUIZ

Refer to the text in this chapter if necessary. A good score is eight correct. Answers are in the back of the book.

1. Figure 12-10 shows a spherical triangle on the surface of the earth, with sides running along the prime meridian, the 67.5° west longitude meridian, and the equator. What's the sum of the measures of the interior angles of this spherical triangle?
   A. $3\pi/4$ rad
   B. π rad
   C. $11\pi/8$ rad
   D. $5\pi/4$ rad

2. What's the distance, to the nearest kilometer, between the meridians for 100.00° west longitude and 105.00° west longitude, as we travel along the parallel corresponding to 30.00° north latitude?
   A. 278 kilometers
   B. 393 kilometers
   C. 482 kilometers
   D. 556 kilometers

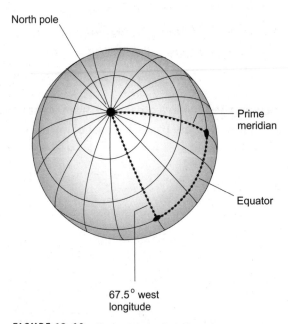

FIGURE 12-10 · Illustration for Quiz Question 1.

3. What's the distance, to the nearest kilometer, between the meridians for 100.00° west longitude and 105.00° west longitude, as we travel along the parallel corresponding to 60.00° north latitude?

   A. 278 kilometers
   B. 393 kilometers
   C. 482 kilometers
   D. 556 kilometers

4. Imagine a spherical triangle whose sides have angular extents of $q = 30°$, $r = 25°$, and $s = 20°$. Also suppose that the spherical angle opposite side q is a right angle. What's the measure of the spherical angle opposite side r to the nearest degree?

   A. We need more information to calculate it.
   B. 69°
   C. 58°
   D. 43°

5. In the situation described by Question 4, what's the measure of the spherical angle opposite side s to the nearest degree?

   A. We need more information to calculate it.
   B. 69°
   C. 58°
   D. 43°

6. What's the sum of the measures of the interior spherical angles of the spherical triangle described by Question 4, to the nearest degree?

   A. 180°
   B. 191°
   C. 164°
   D. 157°

7. Based on the definitions of latitude and longitude given in this chapter, we can deduce that the meridian corresponding to 45° east longitude intersects the equator on the earth's surface at an angle of

   A. $\pi/4$ rad.
   B. $\pi/2$ rad.
   C. π rad.
   D. 2π rad.

8. Figure 12-11 shows a spherical triangle with two sides, having known angular extents (65° and 55°) and a known spherical interior angle (100°) between those two sides. What's the angular extent of the third side to the nearest degree?

   A. 83°
   B. 90°
   C. 108°
   D. 116°

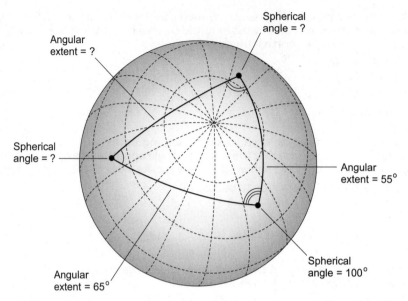

FIGURE 12-11 · Illustration for Quiz Questions 8 through 10.

9. Based on the answer to Question 8 and the situation of Fig. 12-11, what's the measure, to the nearest degree, of the triangle's spherical interior angle at the far-left vertex?

    A. 54°

    B. 64°

    C. 85°

    D. 103°

10. Based on the answer to Question 8 and the situation of Fig. 12-11, what's the measure, to the nearest degree, of the triangle's spherical interior angle at the top-right vertex?

    A. 54°

    B. 64°

    C. 85°

    D. 103°

chapter **13**

The Infinite-Series Paradigm

In trigonometry, we sometimes encounter exotic coincidences. For example, when we add up endless lists of quantities having certain patterns, we can generate all of the circular functions and their inverses. Let's start by learning how, with the help of a little mathematical trickery, we can sometimes add up infinite lists of numbers to get finite sums.

CHAPTER OBJECTIVES

In this chapter, you will

- Learn how to express sequences and series in list form.
- Add up infinite sequences of quantities.
- Compare convergent and divergent series.
- Expand the sine and cosine functions into infinite series.
- Approximate sine and cosine values with partial sums of series.

Repeated Addition

Mathematicians use the term *sequence* to describe a list of numbers. Some sequences are *finite* (they go on for awhile and then come to an end). Other sequences are *infinite* (they go on without end). The simplest sequences have values that repeatedly increase or decrease by a fixed amount. Consider the following examples:

$$A = 1, 2, 3, 4, 5, 6$$

$$B = 0, -1, -2, -3, -4, -5$$

$$C = 2, 4, 6, 8$$

$$D = -5, -10, -15, -20$$

$$E = 4, 8, 12, 16, 20, 24, 28, \ldots$$

$$F = 2, 0, -2, -4, -6, -8, -10, \ldots$$

The first four sequences are finite. The last two are infinite, as indicated by an *ellipsis* (three dots) at the end.

Arithmetic Sequence

In each of the sequences shown above, the values either increase steadily (in A, C, and E) or decrease steadily (in B, D, and F). In all six sequences, the "spacing" between numbers remains constant throughout. Here's how each sequence changes as we move along from term to term:

- The values in A always increase by 1.
- The values in B always decrease by 1.
- The values in C always increase by 2.
- The values in D always decrease by 5.
- The values in E always increase by 4.
- The values in F always decrease by 2.

Each sequence has a specific *initial value*. After that, we can "predict" subsequent values by repeatedly adding a constant. If the added constant is positive, the sequence increases. If the added constant is negative, the sequence decreases.

Suppose that s_0 represents the first number in a sequence S. Let c represent a real-number constant. If we can write S in the form

$$S = s_0, (s_0 + c), (s_0 + 2c), (s_0 + 3c), \ldots$$

then S constitutes an *arithmetic sequence* or an *arithmetic progression*. In this context, we pronounce the word "arithmetic" as "err-ith-MET-ick".

The constants s_0 and c can represent integers, but that's not a requirement. Those quantities can have fractional values such as 2/3 or –7/5, or even irrational values such as π or the square root of 2. As long as the separation between any two adjacent terms remains the same throughout the progression, we have an arithmetic sequence.

Arithmetic Series

When we add up all of the terms in a sequence, we get a *series*. In the case of an arithmetic sequence, we can define the corresponding *arithmetic series* only if the sequence has a finite number of terms. For the foregoing sequences A through F, let's call the corresponding series A_+ through F_+ so that we get total sums as follows:

$$A_+ = 1 + 2 + 3 + 4 + 5 + 6 = 21$$

$$B_+ = 0 + (-1) + (-2) + (-3) + (-4) + (-5) = -15$$

$$C_+ = 2 + 4 + 6 + 8 = 20$$

$$D_+ = (-5) + (-10) + (-15) + (-20) = -50$$

$$\text{We can't define } E_+$$

$$\text{We can't define } F_+$$

PROBLEM **13-1**

Suppose that in an infinite sequence S, we have the first term $s_0 = 5$ and the transitional constant $c = 3$. Calculate the values of the first 10 terms. Then name those terms in the form of a list with an ellipsis at the end.

SOLUTION

The first 10 terms are as follows:

$$s_0 = 5$$

$$s_1 = s_0 + 3$$
$$= 5 + 3$$
$$= 8$$

$$s_2 = s_1 + 3$$
$$= 8 + 3$$
$$= 11$$

$$s_3 = s_2 + 3$$
$$= 11 + 3$$
$$= 14$$

$$s_4 = s_3 + 3$$
$$= 14 + 3$$
$$= 17$$

$$s_5 = s_4 + 3$$
$$= 17 + 3$$
$$= 20$$

$$s_6 = s_5 + 3$$
$$= 20 + 3$$
$$= 23$$

$$s_7 = s_6 + 3$$
$$= 23 + 3$$
$$= 26$$

$$s_8 = s_7 + 3$$
$$= 26 + 3$$
$$= 29$$

$$s_9 = s_8 + 3$$
$$= 29 + 3$$
$$= 32$$

Therefore

$$S = s_0, s_1, s_2, s_3, s_4, s_5, s_6, s_7, s_8, s_9$$
$$= 5, 8, 11, 14, 17, 20, 23, 26, 29, 32, \ldots$$

Still Struggling

You might reasonably ask, "What happens if we start a sequence with a fixed number and then alternately add and subtract a constant? Do we get an arithmetic sequence?" Here's an example:

$$V = -1/2, 1/2, -1/2, 1/2, -1/2, 1/2, -1/2 \ldots$$

In this case, the first term v_0 equals $-1/2$. We might say that the transitional constant c_v equals 1, and we alternately add and subtract it to generate the terms. We can define this sequence, but we can't call it an arithmetic sequence. In order to generate a true arithmetic sequence, we must repeatedly add the transitional constant, whether it's positive, negative, or 0. When the transitional constant is positive, the terms steadily increase. When the transitional constant is negative, the terms steadily decrease. Technically, arithmetic sequences never alternate as V does.

PROBLEM 13-2

When we start to add up the numbers in a sequence, we get another sequence of numbers that we call *partial sums*. List the first five partial sums of the following sequences (not all of which are arithmetic!):

$$S = 5, 8, 11, 14, 17, 20, 23, 26, 29, 32, \ldots$$

$$T = 2, 4, 8, 16, 32, 64, 128, 256, 512, \ldots$$

$$U = 100, 65, 30, -5, -40, -75, -110, \ldots$$

$$V = -1/2, 1/2, -1/2, 1/2, -1/2, 1/2, -1/2 \ldots$$

✔ SOLUTION

We simply add increasing numbers of terms and list the sums. For the sequence S, we get the following values:

$$s_{0+} = 5$$

$$s_{1+} = 5 + 8$$
$$= 13$$

$$s_{2+} = 5 + 8 + 11$$
$$= 24$$

$$s_{3+} = 5 + 8 + 11 + 14$$
$$= 38$$

$$s_{4+} = 5 + 8 + 11 + 14 + 17$$
$$= 55$$

For the sequence T, the first five partial sums are as follows:

$$t_{0+} = 2$$

$$t_{1+} = 2 + 4$$
$$= 6$$

$$t_{2+} = 2 + 4 + 8$$
$$= 14$$

$$t_{3+} = 2 + 4 + 8 + 16$$
$$= 30$$

$$t_{4+} = 2 + 4 + 8 + 16 + 32$$
$$= 62$$

For the sequence U, the first five partial sums are as follows:

$$u_{0+} = 100$$

$$u_{1+} = 100 + 65$$
$$= 165$$

$$u_{2+} = 100 + 65 + 30$$
$$= 195$$

$$u_{3+} = 100 + 65 + 30 + (-5)$$
$$= 190$$

$$u_{4+} = 100 + 65 + 30 + (-5) + (-40)$$
$$= 150$$

For the sequence V, the first five partial sums are as follows:

$$v_{0+} = -1/2$$

$$v_{1+} = -1/2 + 1/2$$
$$= 0$$

$$v_{2+} = -1/2 + 1/2 + (-1/2)$$
$$= -1/2$$

$$v_{3+} = -1/2 + 1/2 + (-1/2) + 1/2$$
$$= 0$$

$$v_{4+} = -1/2 + 1/2 + (-1/2) + 1/2 + (-1/2)$$
$$= -1/2$$

Repeated Multiplication

Another common type of sequence has values that we repeatedly multiply by a transitional constant. Consider the following examples:

$$G = 1, 2, 4, 8, 16, 32$$

$$H = 1, -1, 1, -1, 1, -1, 1 \ldots$$

$$I = 1, 10, 100, 1000$$

$$J = -5, -15, -45, -135, -405$$

$$K = 3, 9, 27, 81, 243, 729, 2187 \ldots$$

$$L = 1/2, 1/4, 1/8, 1/16, 1/32, 1/64, 1/128 \ldots$$

Sequences G, I, and J are finite. Sequences H, K, and L are infinite, as indicated by an ellipsis at the end of each list.

Geometric Sequence

Upon casual observation, the above sequences appear vastly different from one another. But they share one notable similarity: In all six sequences, each term equals a constant multiple of the term immediately before it.

- The values in G progress by a constant factor of 2.
- The values in H progress by a constant factor of –1.
- The values in I progress by a constant factor of 10.
- The values in J progress by a constant factor of 3.
- The values in K progress by a constant factor of 3.
- The values in L progress by a constant factor of $1/2$.

If the constant is positive, the values either remain positive or remain negative. If the constant is negative, the values alternate between positive and negative.

Let's say that t_0 represents the first number in a sequence T. Let k represent a transitional constant. Imagine that we can write T in the general form

$$T = t_0, t_0 k, t_0 k^2, t_0 k^3, t_0 k^4, \ldots$$

for as long as the sequence goes. We call a sequence of this type a *geometric sequence* or *geometric progression*. The numbers t_0 and k can be whole numbers, but they don't have to be. As long as the multiplication factor between any two adjacent terms remains the same, we have a geometric sequence. In the sequence L above, we have the constant $k = 1/2$. This situation gives rise to an especially interesting case, as we'll see in a moment.

Geometric Series

In a geometric sequence, we can always define the corresponding *geometric series*—the sum of all terms—if the sequence is finite. That fact should come as no surprise. However, we can sometimes define the sum of all the terms even if the geometric sequence goes on without end. That fact amazes some people.

For the above sequences G through L, let's call the corresponding series G_+ through L_+. Then we have

$$G_+ = 1 + 2 + 4 + 8 + 16 + 32$$
$$= 63$$
$$H_+ = 1 - 1 + 1 - 1 + 1 - 1 + 1 \ldots$$
$$= ?$$

$$I_+ = 1 + 10 + 100 + 1000$$

$$= 1111$$

$$J_+ = -5 - 15 - 45 - 135 - 405$$

$$= -605$$

K_+ "blows up" (it has no finite sum)

$$L_+ = 1/2 + 1/4 + 1/8 + 1/16 + 1/32 + 1/64 + 1/128 \ldots$$

$$= ?$$

The finite series G_+, I_+, and J_+ work out simply enough. The partial sums of H_+ alternate between 0 and 1, but never settle on either of those values. We might feel tempted to say that H_+ has two values, just as certain equations have solution sets containing two roots. But we're looking for a single, identifiable number, not the solution set of an equation. On that basis, we must conclude that H_+ defies definition. The infinite series K_+ goes "out of control." It's an example of a *divergent series*; the values keep getting farther from 0 forever. As for the infinite series L_+, well, that's where things get interesting!

Convergence

For the above sequences H, K, and L, the sequences of partial sums, which we'll denote using asterisk subscripts, proceed as follows:

$$H_* = 1, 0, 1, 0, 1, 0, \ldots$$

$$K_* = 3, 12, 39, 120, 363, 1092, 3279, \ldots$$

$$L_* = 1/2, 3/4, 7/8, 15/16, 31/32, 63/64, 127/128 \ldots$$

The terms in the sequences H_* and K_* don't settle down on anything. But the terms in the sequence denoted by L_* seem to approach 1. They don't "run away" into uncharted territory, and they don't alternate between different values. The partial sums in L_* seem to have a clear "destination", although we can sense that they'll never quite get all the way there.

As things work out in the exotic world of theoretical mathematics, the complete series L_+, representing the sum of the infinite string of numbers in the sequence L, equals precisely 1! We can get an intuitive view of this fact by observing that the partial sums approach 1. As the position in the sequence of partial sums, L_*, gets farther and farther along, the denominators keep doubling, and the numerator always equals 1 less than the denominator. In fact, if we want to find the nth number L_{*n} in the sequence of partial sums L_*, we can calculate it as

$$L_{*n} = (2^n - 1)/2^n$$

As n becomes large, 2^n becomes large a lot faster, and the proportional difference between $2^n - 1$ and 2^n becomes smaller. When n reaches extremely large positive integer values, the quotient $(2^n - 1)/2^n$ equals almost exactly 1. We can make the quotient as close to 1 as we want by going out far enough in the series of partial sums, but we can never make it equal to 1, and we can certainly never make it bigger than 1. We say that the sequence L_* *converges* on the number 1. The sequence of partial sums L_* therefore provides us with an example of a *convergent sequence*. We call the series L_+ a *convergent series*.

PROBLEM 13-3

Suppose that you buy a five-year certificate of deposit (CD) at your local bank for $1000.00, and it earns interest at the annualized rate of exactly 2% per year. How much will it be worth after six years?

SOLUTION

The CD will be worth $1104.08 after six years. To calculate the value, you multiply $1000 by 1.02, then you multiply that result by 1.02, and you repeat the process a total of five times, rounding off to the nearest cent (hundredth of a dollar) each time. The resulting numbers form a geometric sequence.

- After 1 year: $1000.00 × 1.02 = $1020.00
- After 2 years: $1020.00 × 1.02 = $1040.40
- After 3 years: $1040.40 × 1.02 = $1061.21
- After 4 years: $1061.21 × 1.02 = $1082.43
- After 5 years: $1082.43 × 1.02 = $1104.08

PROBLEM 13-4

Does the following list of numbers portray a geometric sequence? If so, what are the values t_0 (the starting value) and k (the transitional constant)?

$$T = 3, -6, 12, -24, 48, -96, \ldots$$

SOLUTION

This list does indeed describe a geometric sequence. The numbers change by a transitional constant factor of -2. In this case, $t_0 = 3$ and $k = -2$.

Still Struggling

You might wonder, "If a sequence or series has a clear pattern, can we always categorize it as either arithmetic or geometric?" The answer is no! Consider

$$U = 10, 13, 17, 22, 28, 35, 43, \ldots$$

This sequence shows a pattern, but it's neither arithmetic nor geometric. The difference between the first and second terms equals 3, the difference between the second and third terms equals 4, the difference between the third and fourth terms equals 5, and so on. The difference keeps increasing by 1 for each succeeding pair of terms. This list of values provides us with a simple example of a non-arithmetic, non-geometric sequence that nevertheless has an identifiable pattern as we move along. Pretty soon, we'll encounter some others: the *trigonometric function series*.

Limits

Mathematicians define a *limit* as a fixed quantity that a sequence, series, relation, or function approaches as we vary some parameter. The value of the sequence, series, relation, or function can get arbitrarily close to the limit, but doesn't necessarily get all the way there.

Example

Consider an infinite sequence that starts with 1 and then keeps getting smaller. Let's call the series A. For any positive integer n, the nth term equals $1/n$, so we can write the sequence in list form as

$$A = 1, 1/2, 1/3, 1/4, 1/5, \ldots 1/n, \ldots$$

This list portrays an example of a special type of number list called a *harmonic sequence*. In this case, the values of the terms approach 0 as we keep generating more of them. The 10th term equals $1/10$; the 100th term equals $1/100$; the 1000th term equals $1/1000$. If we choose a tiny but positive real number, we can always find a term in the sequence that's closer to 0 than that number, but we'll never get a term that actually equals 0. We say, "The limit of $1/n$, as n approaches infinity, is 0".

Another Example

Consider a sequence of ratios called B in which the numerators ascend one by one through the set of natural numbers (the whole counting numbers), while every denominator equals the corresponding numerator plus 1. For any positive integer n in this sequence, the nth term equals $(n - 1)/n$, so

$$B = 0/1, \ 1/2, \ 2/3, \ 3/4, \ 4/5, \ \dots \ (n - 1) / n, \ \dots$$

As n becomes extremely large, the numerator $(n - 1)$ gets closer and closer to the denominator, when we consider the difference in proportion to the value of n. Therefore, the limit of $(n - 1)/n$, as n approaches infinity, equals 1.

Still Struggling

By now, you might suspect that any given sequence or series must fall into one or the other of two categories: convergent (meaning that it has a limit) or divergent (meaning that it doesn't). But what can we say if a sequence or series alternates between two numbers endlessly? Consider one of the sequences we saw a little while ago:

$$H = 1, -1, 1, -1, 1, -1, 1 \dots$$

This sequence gives rise to the series

$$H_+ = 1 - 1 + 1 - 1 + 1 - 1 + 1 \dots$$

When we add the terms, we get the sequence of partial sums

$$H_* = 1, 0, 1, 0, 1, 0, 1 \dots$$

You ask, "Do sequences and series of this type have two different limits?" The answer is no! A limit must always constitute a single value that we can specify as a number. In each one of these three cases, we have to say that no limit exists at all. Even so, certain types of alternating sequences do have defined limits. We're about to see some of them! But first, we have one more concept to define.

Factorial

For any particular nonnegative integer n greater than or equal to 1, we define the *factorial* of n (also called n *factorial*) as the product of all positive integers from 1 up to, and including, n. Mathematicians use an exclamation mark (!) to represent the word "factorial." Therefore

$$n! = 1 \times 2 \times 3 \times 4 \times \ldots \times n$$

Let's say that $0! = 1$ by default. We won't attempt to define factorials for negative integers, or for any number that's not an integer.

PROBLEM 13-5

Tabulate the values of the factorial function for $n = 0$ through $n = 10$.

SOLUTION

Table 13-1 lists the results. You can find these values with a calculator that can display a lot of digits. Most personal computers have calculators good enough for this purpose. If you're lucky, your calculator has a single key that you can press to get the factorial straightaway, as long as you input a whole number.

TABLE 13-1. Values of $n!$ for $n = 0$ through $n = 10$.

Value of n	Value of $n!$
0	1
1	1
2	2
3	6
4	24
5	120
6	720
7	5040
8	40,320
9	362,880
10	3,628,800

Expansion of the Sine

For any angle θ expressed in radians where $0 \le \theta \le \pi/2$, you can expand its sine into an alternating infinite series as follows:

$$\sin \theta = \theta - \theta^3/3! + \theta^5/5! - \theta^7/7! + \theta^9/9! - \ldots$$

 PROBLEM 13-6

Use the above formula to expand sin 0.1 rad. Assume that the angle value is mathematically exact as stated. Calculate the first five terms of the series to seven decimal places. Compile a table to show how the series converges to the actual value of the function.

✔ SOLUTION

You can use a calculator, along with the factorials from Table 13-1, to do the arithmetic. Leave in all the extra digits that the calculator provides for each step, making use of the memory function to store complicated numbers. This methodology will ensure that you don't get any cumulative rounding errors. You should get Table 13-2. You only have to reach the third term in the series to obtain an approximation of sin 0.1 rad that's accurate to seven decimal places.

PROBLEM 13-7

Use the above formula to expand sin 0.5 rad. Assume that the angle value is exact. Calculate the first five terms of the series to seven decimal places. Compile a table to show how the series converges to the actual value of the function.

TABLE 13-2 Series expansion of sin 0.1 rad through the first five terms. According to a calculator, sin 0.1 rad = 0.0998334. All figures are rounded off to seven decimal places.

Term number	Value according to series
1	0.1000000
2	0.0998333
3	0.0998334
4	0.0998334
5	0.0998334

TABLE 13-3 Series expansion of sin 0.5 rad through the first five terms. According to a calculator, sin 0.5 rad = 0.4794255. All figures are rounded off to seven decimal places.

Term Number	Value according to Series
1	0.5000000
2	0.4791667
3	0.4794271
4	0.4794255
5	0.4794255

✔ **SOLUTION**

You can use a calculator, taking care to leave in all the extra digits as you proceed, and end up with Table 13-3. This time, you'll obtain an approximation that's accurate to seven decimal places when you reach the fourth term in the series.

PROBLEM 13-8

Use the above formula to expand sin 1 rad. Assume that the angle value is exact. Calculate the first five terms of the series to seven decimal places. Compile a table to show how the series converges to the actual value of the function.

✔ **SOLUTION**

Following the same procedure as you did in the previous two problems, but using the new angle value, you should get Table 13-4. This time, you'll

TABLE 13-4 Series expansion of sin 1 rad through the first five terms. According to a calculator, sin 1 rad = 0.8414710. All figures are rounded off to seven decimal places.

Term Number	Value according to Series
1	1.0000000
2	0.8333333
3	0.8416667
4	0.8414683
5	0.8414710

obtain an approximation that's accurate to seven decimal places when you reach the fifth term in the series.

PROBLEM 13-9

Use the above formula to expand sin ($\pi/2$) rad. Assume that the angle value is exact. Calculate the first five terms of the series to seven decimal places. Compile a table to show how the series converges to the actual value of the function. You can use a scientific calculator to figure out Arccos −1 and then divide by 2 to get a multiple-digit display of the value of $\pi/2$. Make sure that you set your calculator for radians, not degrees.

✔ SOLUTION

Once again, you must carry out the arithmetic (expect some tedium here), leaving in all the extra digits that the calculator provides. You'll get Table 13-5. Even when you reach the fifth term, your result will turn out accurate to only five decimal places.

TIP   As an exercise, carry out the result for Problem 13-9, summarized in Table 13-5, to more terms. How many terms do you need in order to attain seven-decimal-place perfection?

PROBLEM 13-10

Based on the solutions to Problems 13-6 through 13-9, what can you conclude about the quality of the results for series expansions of the sine function, for angles ranging from 0 rad through $\pi/2$ rad?

TABLE 13-5 Series expansion of sin ($\pi/2$) rad through the first five terms. According to a calculator, sin ($\pi/2$) rad = 1.0000000. All figures are rounded off to seven decimal places.

Term number	Value according to series
1	1.5707963
2	0.9248322
3	1.0045249
4	0.9998431
5	1.0000035

 SOLUTION

For any given number of terms, the series does the best job of approximating the sine function for the smallest angles, and gradually loses precision as the angles get larger.

Expansion of the Cosine

For any angle θ expressed in radians where $0 \le \theta \le \pi/2$, you can expand its cosine into an alternating infinite series as follows:

$$\cos \theta = 1 - \theta^2/2! + \theta^4/4! - \theta^6/6! + \theta^8/8! - \ldots$$

 PROBLEM **13-11**

Use the above formula to expand cos 0.1 rad. Assume that the angle value is exact. Calculate the first five terms of the series to seven decimal places. Compile a table to show how the series converges to the actual value of the function.

 SOLUTION

You must follow an arithmetic process similar to the one you used in the sine-function problems, except, of course, with the new formula! You should end up with Table 13-6. The series produces a value for cos 0.1 rad accurate to seven decimal places by the time you get to the third term.

TABLE 13-6 Series expansion of cos 0.1 rad through the first five terms. According to a calculator, cos 0.1 rad = 0.9950042. All figures are rounded off to seven decimal places.

Term number	Value according to series
1	1.0000000
2	0.9950000
3	0.9950042
4	0.9950042
5	0.9950042

PROBLEM 13-12

Use the above formula to expand cos 0.5 rad. Assume that the angle value is exact. Calculate the first five terms of the series to seven decimal places. Compile a table to show how the series converges to the actual value of the function.

✔ SOLUTION

Use your calculator, remembering to leave in all the extra digits for each step in the arithmetic. You'll end up with Table 13-7, deriving an approximation that's accurate to seven decimal places when you reach the fifth term in the series.

PROBLEM 13-13

Use the above formula to expand cos 1 rad. Assume that the angle value is exact. Calculate the first five terms of the series to seven decimal places. Compile a table to show how the series converges to the actual value of the function.

✔ SOLUTION

Following the same procedure as you always have, but starting with 1 rad as the input, you should end up with Fig. 13-8. You'll get to within a "gnat's eyelash" of seven perfect decimal places after five terms, but a tiny error remains.

TABLE 13-7   Series expansion of cos 0.5 rad through the first five terms. According to a calculator, cos 0.5 rad = 0.8775826. All figures are rounded off to seven decimal places.

Term Number	Value according to Series
1	1.0000000
2	0.8750000
3	0.8776042
4	0.8775825
5	0.8775826

TABLE 13-8 Series expansion of cos 1 rad through the first five terms. According to a calculator, cos 1 rad = 0.5403023. All figures are rounded off to seven decimal places.

Term Number	Value according to Series
1	1.0000000
2	0.5000000
3	0.5416667
4	0.5402778
5	0.5403026

PROBLEM 13-14

Use the above formula to expand cos $(\pi/2)$ rad. Assume that the angle value is exact. Calculate the first five terms of the series to seven decimal places. Compile a table to show how the series converges to the actual value of the function.

SOLUTION

When you work out the values in the series to the fifth term, you'll get Table 13-9, ending up with a result that's accurate to only four decimal places.

TIP   *As an exercise, carry out the results for Problems 13-13 and 13-14, summarized in Tables 13-8 and 13-9, to the sixth term. Does that extension go far enough to produce a perfect result to seven decimal places? If it doesn't, try going to the seventh term. Keep on going until you attain perfection to seven decimal places.*

TABLE 13-9 Series expansion of cos $(\pi/2)$ rad through the first five terms. According to a calculator, cos $(\pi/2)$ rad = 0.0000000. All figures are rounded off to seven decimal places.

Term Number	Value according to Series
1	1.0000000
2	−0.2337006
3	0.0199690
4	−0.0008945
5	0.0000247

PROBLEM 13-15

Based on the solutions to Problems 13-11 through 13-14, what can you conclude about the quality of the results for series expansions of the cosine function, for angles ranging from 0 rad through $\pi/2$ rad?

✔ SOLUTION

For any given number of terms, the series approximates the cosine function very well for small angles, and not so well for larger angles. You might also notice that the cosine series doesn't approach the actual value as rapidly, term-by-term, as the sine series does, especially for larger angles. Can you think of a reason for this difference?

Still Struggling

Whenever you use either of the foregoing formulas, you'll have to put up with a certain amount of messy, bothersome arithmetic. Before you even start calculating, however, always remember to express the input angle in radians. If you want to find the series expansion for the sine or the cosine of an angle in degrees, you must convert the angle to radians before you do any arithmetic with the series expansions explained in this chapter. If you forget this simple requirement, you'll encounter confusion and chaos (in addition to messiness and tedium)!

TIP   *If you're good at writing software (which I am not), you might want to program your computer to calculate the sine and cosine series for you, saving you the headaches of tedious arithmetic.*

Expansion of the Tangent

You can generate an expansion of the tangent function by taking the ratio of the sine series to the cosine series, as long as the angle in question is positive and remains less than $\pi/2$ rad (that is, $0 \leq \theta < \pi/2$). Always take your calculations out to the same number of terms for each series as you work the ratios. Divide the two expansions only after you've calculated each one to the number of terms that you want. In other words, don't divide the series terms one at a time; divide the worked-out expansions by each other.

PROBLEM 13-16

Use the above-described methodology to expand tan 0.1 rad. Assume that the angle value is exact. Find the ratios of the respective values in Tables 13-2 and 13-6. Compile a new table to show how the partial-sum ratios converge to the actual value of the function.

☑ SOLUTION

Table 13-10 breaks down the process. You don't have to do much arithmetic because you did most of the hard work when you compiled the tables for the sine and the cosine. As with the sine and cosine expansions for 0.1 rad, you only have to go out to the third term to get a value for tan 0.1 rad that's accurate to seven decimal places.

PROBLEM 13-17

Use the above-described methodology to expand tan 0.5 rad. Assume that the angle value is exact. Find the ratios of the respective values in Tables 13-3 and 13-7. Compile a new table to show how the partial-sum ratios converge to the actual value of the function.

TABLE 13-10 Ratios of series expansions of sin 0.1 rad and cos 0.1 rad, yielding tan 0.1 rad through the first five terms. According to a calculator, tan 0.1 rad = 0.1003347. All figures are rounded off to seven decimal places.

Term Number	Sine Value	Cosine Value	Tangent Value
1	0.1000000	1.0000000	0.1000000
2	0.0998333	0.9950000	0.1003350
3	0.0998334	0.9950042	0.1003347
4	0.0998334	0.9950042	0.1003347
5	0.0998334	0.9950042	0.1003347

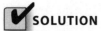 **SOLUTION** _____

If you use the exact figures from Tables 13-3 and 13-7 (rounded off to seven decimal places), you should get the result shown in Table 13-11. Interestingly, you'll get an answer that's accurate to seven decimal places for the ratio of the fourth terms, but when you calculate the ratio of the fifth terms, you'll be off by one digit in the "ten-millions" place!

TIP   *The discrepancy in the lower-rightmost value in Table 13-11 results from the fact that you used rounded-off values to generate the ratios. If you'd taken all of the terms for Tables 13-3 and 13-7 to, say, 12 decimal places instead of only seven, you'd get seven-decimal-place perfection here.*

 PROBLEM 13-18 _____

Use the above-described methodology to expand tan 1 rad. Assume that the angle value is exact. Find the ratios of the respective values in Tables 13-4 and 13-8. Compile a new table to show how the partial-sum ratios converge to the actual value of the function.

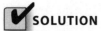 **SOLUTION** _____

Table 13-12 shows the result. As you'll probably expect, the ratios of the series terms produce a new converging sequence, but it doesn't approach the actual value quite as fast as the ratios did for smaller angles. Even when you get to the fifth term, you still have a small error.

TABLE 13-11 Ratios of series expansions of sin 0.5 rad and cos 0.5 rad, yielding tan 0.5 rad through the first five terms. According to a calculator, tan 0.5 rad = 0.5463025. All figures are rounded off to seven decimal places.

Term Number	Sine Value	Cosine Value	Tangent Value
1	0.5000000	1.0000000	0.5000000
2	0.4791667	0.8750000	0.5476191
3	0.4794271	0.8776042	0.5462908
4	0.4794255	0.8775825	0.5463025
5	0.4794255	0.8775826	0.5463024

TABLE 13-12 Ratios of series expansions of sin 1 rad and cos 1 rad, yielding tan 1 rad through the first five terms. According to a calculator, tan 1 rad = 1.5574077. All figures are rounded off to seven decimal places.

Term Number	Sine Value	Cosine Value	Tangent Value
1	1.0000000	1.0000000	1.0000000
2	0.8333333	0.5000000	1.6666667
3	0.8416667	0.5416667	1.5538461
4	0.8414683	0.5402778	1.5574734
5	0.8414710	0.5403026	1.5574069

PROBLEM 13-19

Use the above-described methodology to expand tan $(\pi/2)$ rad. Assume that the angle value is exact. Find the ratios of the respective values in Tables 13-5 and 13-9. Compile a new table to show how the partial-sum ratios "blow up." Remember that the actual value of tan $(\pi/2)$ is undefined.

SOLUTION

You'll get Table 13-13. Can you see what's happening here? The sequence of term ratios diverges in both directions, positive and negative. If you want to carry the process further, you'll see a continuation of this positive/negative "dual divergence."

TABLE 13-13 Expansion ratios for sin $(\pi/2)$ rad and cos $(\pi/2)$ rad, yielding tan $(\pi/2)$ rad through the first five terms. As you should know by now, tan $(\pi/2)$ rad is undefined! All figures in this table are nevertheless rounded off to seven decimal places.

Term Number	Sine Value	Cosine Value	Tangent Value
1	1.5707963	1.0000000	1.5707963
2	0.9248322	−0.2337006	−3.9573377
3	1.0045249	0.0199690	50.3042165
4	0.9998431	−0.0008945	−1117.7675797
5	1.0000035	0.0000247	40,485.9716599

Still Struggling

Do you feel tempted to say that, based on the divergence of the term-ratio sequence in Table 13-13, tan ($\pi/2$) rad equals "positive and negative infinity at the same time"? If you look back at Fig. 3-5 in Chap. 3 (the graph of the tangent function), you'll get some reinforcement for that idea. Resist it! The value of tan ($\pi/2$) rad must remain undefined unless you can invent a mathematically rigorous way to define it. You'll also have to define "positive infinity" and "negative infinity." If you'd like to attempt such definitions, go ahead! I had a lot of fun with "infinity" from middle school all the way through college. If you enjoy pure but exotic mathematics, I suggest that you explore the mystery of tan ($\pi/2$). If you solve it, please let me know!

PROBLEM 13-20

What can you conclude from the solutions to Problems 13-16 through 13-19, concerning the sequence of term ratios for the tangent function over the range of angles from 0 rad through $\pi/2$ rad?

✔ SOLUTION

The values of the sine and cosine series terms approach the actual function values more and more slowly as the angle increases from 0 rad to $\pi/2$ rad. You'll see that the term ratios do the same thing for the tangent function until you get extremely close to $\pi/2$ rad. Then the ratio "blows up."

TIP   *You can create an expansion of the cosecant function by taking the reciprocal of the sine series, an expansion of the secant function by taking the reciprocal of the cosine series, and an expansion of the cotangent function by taking the ratio of the cosine series to the sine series. When you work with a ratio of the cosine series to the sine series, always go out to the same number of terms for each series.*

TIP   *For any angle in the range 0 rad to $\pi/2$ rad, the expansion for a circular function converges if and only if you can define the function for that angle. If you can't define the function for a particular angle, then no convergent expansion exists for that angle.*

QUIZ

Refer to the text in this chapter if necessary. A good score is eight correct. Answers are in the back of the book.

1. **In an arithmetic sequence, the difference between any given term and the next one**
   A. remains constant as we move along.
   B. keeps getting smaller as we move along.
   C. keeps getting larger as we move along.
   D. alternates between positive and negative values as we move along.

2. **The sum of all the terms in an infinite geometric sequence is**
   A. always finite.
   B. always infinite.
   C. sometimes finite and sometimes infinite.
   D. never defined.

3. **Consider an infinite sequence F in which the nth term equals the reciprocal of $n!$, as follows:**

$$F = 1/1!, 1/2!, 1/3!, 1/4!, 1/5!, \ldots, 1/n!, \ldots$$

   As we go out in this sequence indefinitely, the term value
   A. grows without bound.
   B. approaches n.
   C. approaches $(n + 1)/n$.
   D. approaches 0.

4. **Consider an infinite sequence G in which the nth term equals the factorial of $n - 1$ divided by the factorial of n, as follows:**

$$G = 0!/1!, 1!/2!, 2!/3!, 3!/4!, 4!/5!, \ldots, (n-1)!/n!, \ldots$$

   As we go out in this sequence indefinitely, the term value
   A. grows without bound.
   B. approaches $n - 1$.
   C. approaches $(n + 1)/n$.
   D. approaches 0.

5. **When an infinite series converges, we can identify**
   A. a single limit value.
   B. no single limit value.
   C. any finite number of limit values.
   D. infinitely many limit values.

6. As we go out indefinitely in the series expansion for sin π/4 rad, the term value
   A. grows without bound.
   B. approaches 0.
   C. approaches π/4.
   D. approaches 1.

7. As we go out indefinitely in the series expansion for sin 1 rad, the term value
   A. grows without bound.
   B. approaches 0.
   C. approaches π/4.
   D. approaches 1.

8. In the range of angles θ in radians from 0 to π/2, the infinite series expansion for the sine function converges when
   A. $0 < \theta < \pi/2$.
   B. $0 \le \theta < \pi/2$.
   C. $0 < \theta \le \pi/2$.
   D. All of the above

9. In the range of angles θ in radians from 0 to π/2, the infinite series expansion for the cosine function converges when
   A. $0 < \theta < \pi/2$.
   B. $0 \le \theta < \pi/2$.
   C. $0 < \theta \le \pi/2$.
   D. All of the above

10. In the range of angles θ in radians from 0 to π/2, the infinite series expansion for the tangent function converges when
    A. $0 < \theta < \pi/2$.
    B. $0 \le \theta < \pi/2$.
    C. $0 < \theta \le \pi/2$.
    D. All of the above

Test: Part II

Do not refer to the text when taking this test. You may draw diagrams or use a calculator if necessary. A good score is at least 38 correct. Answers are in the back of the book. It's best to have a friend check your score the first time, so you won't memorize the answers if you want to take the test again.

1. In an AC sine wave, π rad of phase corresponds to exactly
   A. 1/8 of a cycle.
   B. 1/4 of a cycle.
   C. 1/2 of a cycle.
   D. 1 cycle.
   E. 2 cycles.

2. When parallel rays of light strike a convex spherical mirror, the reflected rays
   A. emerge parallel to each other.
   B. converge toward a point.
   C. diverge.
   D. emerge in a random way.
   E. emerge perpendicular to the mirror surface at every point.

3. When I use my computer's calculator program to display the cosine of 89.9999°, it shows me

   1.7453292519934434807679896054328e−6

   Rounded off to three significant figures in scientific notation, this expression works out as
   A. 1.745×10^{-6}
   B. 1.75×10^{-6}
   C. 1.745×10^{6}
   D. 0.000174
   E. 174,000

4. When the terms of an infinite series approach a specific, finite value as we keep adding up partial sums, we say that the series
   A. converges.
   B. diverges.
   C. asymptotes.
   D. exceeds a limit.
   E. has no limit.

5. How tall is the person shown in Fig. Test II-1?
   A. 1.891 m
   B. 1.882 m
   C. 1.877 m
   D. 1.861 m
   E. 1.855 m

6. In the range of angles θ in radians from 0 to $\pi/2$, the infinite series expansion for the cotangent function converges if and only if
   A. $0 < \theta < \pi/2$.
   B. $0 \le \theta < \pi/2$.
   C. $0 < \theta \le \pi/2$.
   D. $0 \le \theta \le \pi/2$.
   E. None of the above

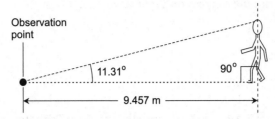

Observation point

11.31°

90°

9.457 m

FIGURE TEST II-1 · Illustration for Part II Test Question 5.

7. **The accuracy in meters with which a surveyor can measure the distance to an object depends on**

   A. the length of the observation base line.
   B. the actual distance to the object.
   C. the time of day.
   D. the angle of the sun above the horizon.
   E. More than one of the above

8. **If we extend a geodesic arc to form a full circle around the earth (assuming that the earth defines a "perfectly round" sphere), then the center of that circle corresponds to**

   A. a point on the earth's surface.
   B. a point inside the earth that depends on the latitude of the geodesic.
   C. a point inside the earth that depends on the radius of the geodesic.
   D. a point outside the earth.
   E. the earth's center.

9. **What's the period of an AC wave whose frequency equals 40,000 Hz?**

   A. 2.5000×10^{-5} s
   B. 2.5000×10^{-6} s
   C. 5.0000×10^{-7} s
   D. 4.0000×10^{-7} s
   E. 3.3333×10^{-7} s

10. **For the wave described in Question 9, how much time would represent exactly $\pi/10$ rad of phase?**

    A. 1.2500×10^{-5} s
    B. 2.5000×10^{-6} s
    C. 1.2500×10^{-6} s
    D. 5.0000×10^{-7} s
    E. 2.0000×10^{-7} s

11. **We can also write the expression $\text{Tan}^{-1} \theta$ as**

    A. $1/(\tan \theta)$
    B. $(\tan \theta)^{-1}$
    C. $\text{Arctan } \theta$
    D. $\tan(1/\theta)$
    E. Any of the above

12. We can also write the expression $\tan^2 \theta$ as
   A. $1/(\tan \theta)^2$
   B. $(\tan \theta)^2$
   C. $\tan(\theta^2)$
   D. $1/\tan(\theta^2)$
   E. None of the above

13. In the situation of Fig. Test II-2, suppose that $b = 1.57000$ meters and $\theta = 25.0000°$, both accurate to six significant figures. What's the distance x to the nearest hundredth of a meter?
   A. 3.37 meters
   B. 3.71 meters
   C. 4.02 meters
   D. 4.18 meters
   E. We need more information to calculate it.

14. In the situation of Question 13 and Fig. Test II-2, what's the distance y to the nearest hundredth of a meter?
   A. 3.37 meters
   B. 3.71 meters
   C. 4.02 meters
   D. 4.18 meters
   E. We need more information to calculate it.

15. In the situation of Fig. Test II-2, suppose that b equals *exactly* 1 meter (accurate to as many significant digits as we want) and θ equals *exactly* 1 second of arc (again, accurate to as many significant digits as we want). What's the distance x to the nearest thousandth of a meter?
   A. 206,264.803 meters
   B. 206,264.804 meters
   C. 206,264.805 meters
   D. 206,264.806 meters
   E. 206,264.807 meters

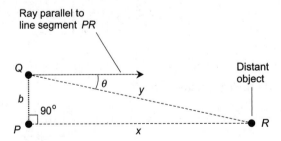

FIGURE TEST II-2 · Illustration for Part II Test Questions 13 through 16.

16. In the situation of Question 15 and Fig. Test II-2, what's the distance y to the nearest thousandth of a meter?
   A. 206,264.803 m
   B. 206,264.804 m
   C. 206,264.805 m
   D. 206,264.806 m
   E. 206,264.807 m

17. When parallel rays of light strike a concave spherical mirror, the reflected rays
   A. emerge parallel to each other.
   B. converge toward a point.
   C. diverge.
   D. emerge in a random way.
   E. emerge perpendicular to the mirror surface at every point.

18. What do engineers call the length of time between a specific point on an AC wave and the same point on the next repetition of the wave?
   A. The cycle
   B. The frequency
   C. The wavelength
   D. The hertz
   E. The period

19. Figure Test II-3 shows how a concave mirror forms a real image at a focal point. Suppose that the initial distance r_o between the mirror and the object equals 110 times the mirror's focal length f. If we quadruple r_o to $440f$, what happens to the distance r_i between the mirror and the object's image?
   A. It doesn't change at all.
   B. It increases slightly.
   C. It increases considerably.
   D. It decreases slightly.
   E. It decreases considerably.

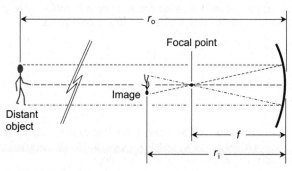

FIGURE TEST II-3 · Illustration for Part II Test Questions 19 and 20.

20. In the situation of Question 19 and Fig. Test II-3, suppose that the initial distance r_o between the mirror and the object equals 1.1 times the mirror's focal length f. If we quadruple r_o to $4.4f$, what happens to the distance r_i between the mirror and the object's image?

    A. It doesn't change at all.
    B. It increases slightly.
    C. It increases considerably.
    D. It decreases slightly.
    E. It decreases considerably.

21. What's the difference $(2.54 \times 10^{-5}) - (7.73 \times 10^{12})$ taking significant figures into account?

    A. 7.73×10^{12}
    B. 7.73×10^{-12}
    C. -7.73×10^{12}
    D. 2.54×10^{-5}
    E. -2.54×10^{-5}

22. Suppose that we draw a triangle on the surface of a sphere so that the triangle's vertices all lie along a single geodesic. In that case, the perimeter of the triangle equals

    A. $2/\pi$ times the sphere's circumference.
    B. half the sphere's circumference.
    C. $1/\pi$ times the sphere's circumference.
    D. $\pi/4$ times the sphere's circumference.
    E. the sphere's circumference.

23. If we measure the interior angles of any triangle such as Question 22 describes, we'll find that the sum of those angle measures equals

    A. 3π rad.
    B. 2π rad.
    C. π rad.
    D. $\pi/2$ rad.
    E. $\pi/3$ rad.

24. If two AC sine waves have the same frequency but differ in phase by 1/10 of a cycle, we might also say that the phase difference equals

    A. $\pi/10$ rad.
    B. $\pi/5$ rad.
    C. $2\pi/5$ rad.
    D. $\pi/20$ rad.
    E. $\pi/36$ rad.

25. According to a calculator, what's $\tan(\pi/6)$ rounded off to three significant figures? (To obtain a value for π, use your calculator to determine the Arccosine of -1, making sure to set the calculator for radians.)

    A. 0.577
    B. 0.500

C. 0.866
D. 1.00
E. It's not defined.

26. **Suppose that we observe two stars: Star A exactly 10 parsecs away from us, and Star B exactly 40 parsecs away from us. From this information, we know that**

A. Star B is 16 times as far away as Star A.
B. Star B is four times as far away as Star A.
C. Star B is twice as far away as Star A.
D. Star A is twice as far away as Star B.
E. Star A is four times as far away as Star B.

27. **Figure Test II-4 shows a ray of light passing through a flat boundary from an upper medium with a refractive index of 1.20 to a lower medium with a refractive index of 1.70. According to Snell's law, the sine of the lower angle ϕ divided by the sine of the upper angle of 57° equals**

A. $1.20/1.70$.
B. $(1.20/1.70)^2$.
C. Arcsin $(1.20/1.70)$.
D. $1.70/1.20$.
E. Arctan $(1.70/1.20)$.

28. **In the situation of Fig. Test II-4, angle ϕ measures**

A. 24°.
B. 30°.
C. 33°.
D. 36°.
E. 41°.

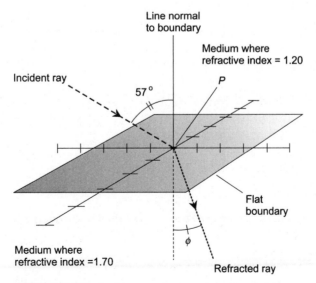

FIGURE TEST II-4 · Illustration for Part II Test Questions 27 and 28.

29. According to a calculator, what's the value of cos 0°0'17", accurate to five significant figures? (Assume that the angle value is mathematically exact, and remember to *round off* the final result!)

A. 0.99999

B. 1.0000

C. 0.00001

D. 0.00000

E. It's not defined.

30. What's the conventional frequency of a wave whose angular frequency equals 18,849,556 rad/s, expressed to three significant figures?

A. 500 kHz

B. 1.00 MHz

C. 2.00 MHz

D. 3.00 MHz

E. 6.00 MHz

31. As we go out indefinitely in the series expansion for cos $\pi/3$ rad, the term value

A. grows without bound.

B. approaches 0.

C. approaches 1/2.

D. approaches 1/3.

E. approaches $1/\pi$.

32. In the range of angles θ in radians from 0 to $\pi/2$, the infinite series expansion for the secant function converges if and only if

A. $0 < \theta < \pi/2$.

B. $0 \le \theta < \pi/2$.

C. $0 < \theta \le \pi/2$.

D. $0 \le \theta \le \pi/2$.

E. None of the above

33. Which of the following infinite series converges?

A. $1 + 2 + 3 + 4 + 5 + 6 + 7 \cdots$

B. $4 + 3 + 2 + 1 + 0 + (-1) + (-2) + \cdots$

C. $1 - 1 + 1 - 1 + 1 - 1 + 1 +$

D. $1 - 1/2! + 1/4! - 1/6! + 1/8! - 1/10! + 1/12!$

E. All of the above

34. What's the angular frequency of a wave whose period equals 2.0000×10^{-4} s?

A. 1.5708×10^4 rad/s

B. 3.1416×10^4 rad/s

C. 7.8540×10^4 rad/s

D. 2.6180×10^5 rad/s

E. 5.2360×10^5 rad/s

35. **When parallel rays of light strike a flat mirror, the reflected rays**
    A. emerge parallel to each other.
    B. converge toward a point.
    C. diverge.
    D. emerge in a random way.
    E. emerge perpendicular to the mirror surface at every point.

36. **In the range of angles θ in radians from 0 to $\pi/2$, the infinite series expansion for the cosecant function converges if and only if**
    A. $0 < \theta < \pi/2$.
    B. $0 \le \theta < \pi/2$.
    C. $0 < \theta \le \pi/2$.
    D. $0 \le \theta \le \pi/2$.
    E. None of the above

37. **Figure Test II-5 shows a triangle on the surface of a "perfectly round" sphere. The measures of the interior angles all equal ψ. The angular extents of the sides equal q, r, and s as shown. We can have *absolute confidence* that**
    A. $q = r$
    B. $r = s$
    C. $q = s$
    D. All of the above
    E. None of the above

38. **In the situation of Fig. Test II-5, we can have *absolute confidence* that**
    A. $\psi = \pi/6$
    B. $\psi = \pi/4$
    C. $\psi = \pi/3$
    D. $\psi = \pi/2$
    E. None of the above

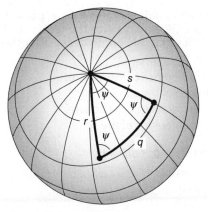

FIGURE TEST II-5 · Illustration for Part II
Test Questions 37 and 38.

39. As we go out indefinitely in the series expansion for sin 30°, the term value

   A. grows without bound.
   B. approaches 0.
   C. approaches 1/2.
   D. approaches 1/3.
   E. approaches 1/π.

40. Suppose that we draw a pentagon on the surface of a sphere in such a way that all five of the pentagon's vertices lie along a single geodesic. In that case, the perimeter of the pentagon equals

   A. 5/π times the sphere's circumference.
   B. twice the sphere's circumference.
   C. 2/π times the sphere's circumference.
   D. π/5 times the sphere's circumference.
   E. the sphere's circumference.

41. If we measure the interior angles of any pentagon such as Question 40 describes, we'll find that the sum of those angle measures equals

   A. 630°.
   B. 720°.
   C. 810°.
   D. 900°.
   E. 1080°.

42. According to a calculator, what's Arccos 0.5000 in degrees, rounded off to four significant figures?

   A. 26.57°
   B. 30.00°
   C. 63.43°
   D. 60.00°
   E. It's not defined.

43. Figure Test II-6 shows two points Q and R on a "perfectly round" sphere. If the length of line segment QR equals twice the radius of the sphere, then we can have *absolute confidence* that the length of arc QR equals

   A. the circumference of the sphere.
   B. 2/π times the circumference of the sphere.
   C. half the circumference of the sphere.
   D. π/8 times the circumference of the sphere.
   E. π/12 times the circumference of the sphere.

44. The quantities 10,000 and 10^{-3} differ by

   A. three orders of magnitude.
   B. four orders of magnitude.
   C. 12 orders of magnitude.
   D. 16 orders of magnitude.
   E. None of the above

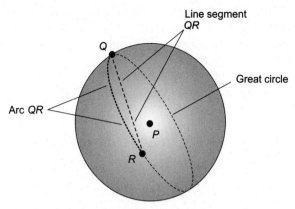

Line segment
QR

Q

Great circle

Arc QR

P

R

FIGURE TEST II-6· Illustration for Part II Test Question 43.

45. **When two sine waves of the same frequency exist in phase quadrature, one of the waves leads or lags the other by**

   A. 90°.
   B. $\pi/4$ rad.
   C. 4°.
   D. 4π rad.
   E. 180°.

46. **When a ray of light crosses a flat boundary from a clear medium having a certain refractive index to another clear medium having a lower refractive index, the measure of the critical angle equals the Arcsine of the ratio of the smaller index to the larger index. Imagine that we shine a laser beam up from the bottom of a tank full of liquid whose refractive index equals 1.23. Above the liquid, the atmosphere has a refractive index of 1.00. Suppose that the light beam intersects the surface at an angle of incidence θ with respect to the normal. How large, at a minimum, must we make θ in order to observe total internal reflection, so that the laser beam stays in the liquid and never enters the air above it?**

   A. The premise of this question is wrong. Under these circumstances, total internal reflection will not occur at any angle!
   B. 54.4°
   C. 60.0°
   D. 63.3°
   E. 66.7°

47. **What do engineers call the number of complete pattern repetitions per second in an AC wave?**

   A. The cycle
   B. The frequency
   C. The wavelength
   D. The hertz
   E. The period

48. For any triangle that lies in a flat plane, the lengths of the sides exist in a constant ratio relative to the
    A. measures of the angles opposite those sides.
    B. cosines of the angles opposite those sides.
    C. sines of the angles opposite those sides.
    D. tangents of the angles opposite those sides.
    E. reciprocals of the angles opposite those sides.

49. In Fig. Test II-7A, the "circle that lies in a plane parallel to the plane of the equator," as illustrated, represents

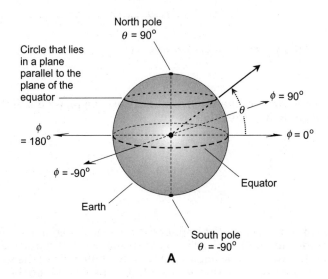

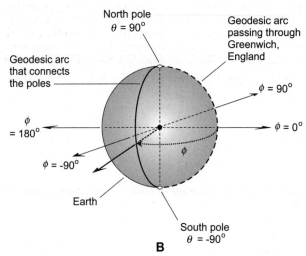

FIGURE TEST II-7 · Illustration for Part II Test Questions 49 and 50.

A. the meridian whose west longitude equals θ.
B. the parallel whose north latitude equals θ.
C. a geodesic whose north latitude equals θ.
D. the plane that defines $1/\pi$ of the sphere's volume.
E. the circle corresponding to $\theta = $ Arcsin (1/2).

50. **In Fig. Test II-7B, the "geodesic arc that connects the poles," as illustrated, represents**

A. the meridian whose west longitude equals ϕ.
B. the parallel whose north latitude equals ϕ.
C. a geodesic whose north latitude equals ϕ.
D. the plane that defines $1/\pi$ of the sphere's volume.
E. the circle corresponding to $\phi = $ Arcsin (1/2).

Final Exam

Do not refer to the text when taking this test. To solve some of these problems, you'll need a good scientific calculator that can carry operations out to a lot of decimal places. A good score is at least 75 correct. The correct answers are listed in the back of the book. It's best to have a friend check your score the first time, so you won't memorize the answers if you want to take the test again.

1. Imagine a perfectly vertical flag pole in the middle of a vast, flat, level field on a sunny day. The pole measures exactly 20 meters in height, and casts a shadow that's exactly 48 meters long. How long, rounded to the nearest meter, is the *sun line* (the straight line segment that goes through the air from the top of the flagpole to the end of its shadow)?

   A. 52 meters
   B. 54 meters
   C. 56 meters
   D. 59 meters
   E. 63 meters

2. In the scenario of Question 1, what's the cosine of the angle between the flag-pole and the sun line?

   A. The pole height divided by the shadow length
   B. The shadow length divided by the sun-line length
   C. The pole height divided by the sun-line length
   D. The shadow length divided by the pole height
   E. None of the above

3. In the scenario of Question 1, what's the sine of the angle between the flagpole and the sun line?

   A. The pole height divided by the shadow length
   B. The shadow length divided by the sun-line length
   C. The pole height divided by the sun-line length
   D. The shadow length divided by the pole height
   E. None of the above

4. In the scenario of Question 1, what's the tangent of the angle between the flag-pole and the sun line?

   A. The pole height divided by the shadow length
   B. The shadow length divided by the sun-line length
   C. The pole height divided by the sun-line length
   D. The shadow length divided by the pole height
   E. None of the above

5. In the scenario of Question 1, what's the cotangent of the angle between the flagpole and the sun line?

   A. The pole height divided by the shadow length
   B. The shadow length divided by the sun-line length
   C. The pole height divided by the sun-line length
   D. The shadow length divided by the pole height
   E. None of the above

6. In the scenario of Question 1, what's the cosecant of the angle between the flagpole and the sun line?

   A. The sun-line length divided by the shadow length
   B. The shadow length divided by the pole height

C. The pole height divided by the shadow length
D. The sun-line length divided by the pole height
E. None of the above

7. **In the scenario of Question 1, what's the secant of the angle between the flag-pole and the sun line?**
   A. The sun-line length divided by the shadow length
   B. The shadow length divided by the pole height
   C. The pole height divided by the shadow length
   D. The sun-line length divided by the pole height
   E. None of the above

8. **In the scenario of Question 1, what's the Arctangent of the angle between the flagpole and the sun line?**
   A. The sun-line length divided by the shadow length
   B. The shadow length divided by the pole height
   C. The pole height divided by the shadow length
   D. The sun-line length divided by the pole height
   E. None of the above

9. **In Fig. Exam-1, how far does point _P_ lie from the origin?**
   A. The square root of 7 units
   B. The square root of 12 units
   C. 5 units
   D. 7 units
   E. We need more information to say.

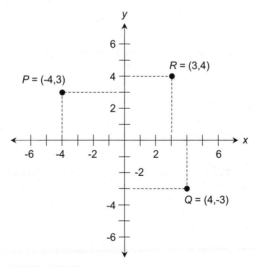

FIGURE EXAM-1 • Illustration for Final Exam
Questions 9 through 15.

10. **In Fig. Exam-1, how far does point Q lie from the origin?**
    A. The square root of 7 units
    B. The square root of 12 units
    C. 5 units
    D. 7 units
    E. We need more information to say.

11. **In Fig. Exam-1, how far does point P lie from point Q?**
    A. 10 units
    B. The square root of 50 units
    C. The square root of 37 units
    D. 14 units
    E. We need more information to say.

12. **In Fig. Exam-1, what are the coordinates of the midpoint of the line segment connecting points P and Q?**
    A. (1/2,1/2)
    B. (−1/2,1/2)
    C. (1/2,−1/2)
    D. (−1/2,−1/2)
    E. None of the above

13. **In Fig. Exam-1, how far does point R lie from the origin?**
    A. The square root of 7 units
    B. The square root of 12 units
    C. 5 units
    D. 7 units
    E. We need more information to say.

14. **In Fig. Exam-1, how far does point R lie from point Q?**
    A. 10 units
    B. The square root of 50 units
    C. The square root of 37 units
    D. 14 units
    E. We need more information to say.

15. **In Fig. Exam-1, what are the coordinates of the midpoint of the line segment connecting points R and Q?**
    A. (7/2,1/2)
    B. (7/2,1/3)
    C. (7/2,1/4)
    D. (7/2,1/5)
    E. None of the above

16. The value of tan θ is undefined in the unit-circle paradigm if and only if the abscissa (independent-variable coordinate) equals

   A. 0.
   B. 1 or −1.
   C. $2^{1/2}/2$ or $−2^{1/2}$.
   D. 2 or −2.
   E. ∞ or −∞.

17. The value of cot θ is undefined in the unit-circle paradigm if and only if the abscissa equals

   A. 0.
   B. 1 or −1.
   C. $2^{1/2}/2$ or $−2^{1/2}$.
   D. 2 or −2.
   E. ∞ or −∞.

18. Figure Exam-2 shows a unit circle with an angle θ inside it, along with two displacements, s (horizontal) and t (vertical). What does s represent?

   A. csc θ
   B. sec θ
   C. tan θ
   D. cos θ
   E. None of the above

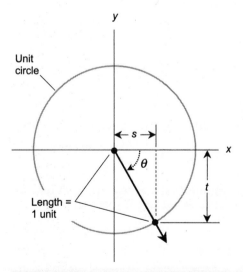

FIGURE EXAM-2 · Illustration for Final Exam Questions 18 through 23.

19. **In Fig. Exam-2, what does *t* represent?**
    A. csc θ
    B. sec θ
    C. tan θ
    D. cos θ
    E. None of the above

20. **In Fig. Exam-2, what does 1/*s* represent?**
    A. csc θ
    B. sec θ
    C. tan θ
    D. cos θ
    E. None of the above

21. **In Fig. Exam-2, what does 1/*t* represent?**
    A. csc θ
    B. sec θ
    C. tan θ
    D. cos θ
    E. None of the above

22. **In Fig. Exam-2, what does *s*/*t* represent?**
    A. csc θ
    B. sec θ
    C. tan θ
    D. cos θ
    E. None of the above

23. **In Fig. Exam-2, what does *t*/*s* represent?**
    A. csc θ
    B. sec θ
    C. tan θ
    D. cos θ
    E. None of the above

24. **When we see a mapping that creates a perfect one-to-one correspondence between two sets of numbers, we call that mapping**
    A. an injection.
    B. a rejection.
    C. a monojection.
    D. a bijection.
    E. a surjection.

25. Which of the following ordered pairs in the form (x, y) does *not* belong to the function $y = \sin x$ when we express x in radians?

    A. $(0, 0)$
    B. $(\pi/6, 1/2)$
    C. $(\pi/3, 1)$
    D. $(3\pi/2, -1)$
    E. $(\pi, 0)$

26. Figure Exam-3 shows a point (x_0, y_0) on the unit circle in the first quadrant of the Cartesian xy-plane. A ray from the origin through (x_0, y_0) subtends an angle θ, which we express in radians relative to the positive x axis. We get θ when we execute *one and only one* of the following functions. Which one?

    A. Arccos $1/x_0$
    B. Arccos x_0
    C. Arcsin x_0
    D. Arctan y_0
    E. Arccsc y_0

27. In the scenario of Fig. Exam-3, we get θ when we execute *one and only one* of the following functions. Which one?

    A. Arccsc $1/y_0$
    B. Arccot x_0
    C. Arccot y_0/x_0
    D. Arcsec y_0
    E. Arccos $1/y_0$

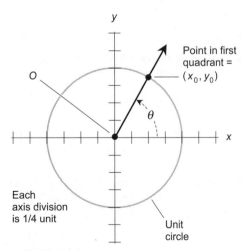

FIGURE EXAM-3 • Illustration for Final Exam Questions 26 through 28.

28. In the scenario of Fig. Exam-3, we get θ when we execute *any* of the following functions *except one*. Which one?

    A. Arcsec $1/x_0$
    B. Arccot x_0/y_0
    C. Arctan y_0/x_0
    D. Arccsc x_0
    E. Arcsin y_0

29. Why do we generally restrict the values of the inverse circular functions (such as the Arcsine or the Arccosine) to a principal branch?

    A. To keep the functions from becoming negative at any point.
    B. To make sure that they behave as true functions.
    C. To keep the function values within the closed interval $[-1,1]$.
    D. All of the above
    E. The premise is wrong! We need not, and in fact should not, restrict the values of the inverse circular functions in any way.

30. The expression $(\tan x)^{-1}$ means

    A. The Arctangent of x.
    B. The tangent of $1/x$.
    C. The reciprocal of the tangent of x.
    D. The cotangent of x.
    E. The Arccotangent of x.

31. The expression $\mathrm{Tan}^{-1} x$ means

    A. The Arctangent of x.
    B. The tangent of $1/x$.
    C. The reciprocal of the tangent of x.
    D. The cotangent of x.
    E. The Arccotangent of x.

32. Figure Exam-4 is a graph of the unit hyperbola in a Cartesian plane. What's the equation of the complete curve?

    A. $u^2 + v^2 = 1$
    B. $u^2 - v^2 = 1$
    C. $u = v^2$
    D. $v = u^2$
    E. $u^2 = v^3$

33. In Fig. Exam-4, the point P has the coordinates (u,v). What's the hyperbolic sine of x ($\sinh x$), where x equals twice the area of the shaded region?

    A. u
    B. v
    C. u/v
    D. v/u
    E. $1/u$

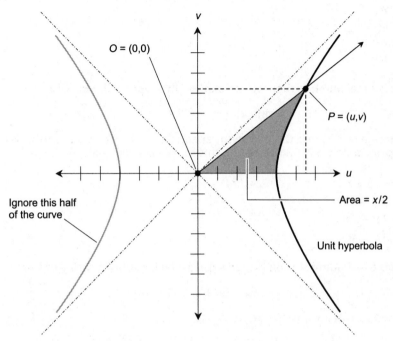

Each increment on the *u* axis equals 1/4 unit
and
each increment on the *v* axis equals 1/4 unit

FIGURE EXAM-4 • Illustration for Final Exam Questions 32 through 36.

34. **As shown in Fig. Exam-4, what's the hyperbolic cosine of *x* (cosh *x*), where *x* equals twice the area of the shaded region?**

    A. *u*
    B. *v*
    C. *u/v*
    D. *v/u*
    E. *1/u*

35. **As shown in Fig. Exam-4, what's the hyperbolic tangent of *x* (tanh *x*), where *x* equals twice the area of the shaded region?**

    A. *u*
    B. *v*
    C. *u/v*
    D. *v/u*
    E. *1/u*

36. As shown in Fig. Exam-4, what's the hyperbolic cotangent of x (coth x), where x equals twice the area of the shaded region?

   A. u
   B. v
   C. u/v
   D. v/u
   E. $1/u$

37. If x is a real number, then we can express the hyperbolic sine of x as

   $$\sinh x = (e^x - e^{-x})/2$$

   where e represents the exponential constant, an irrational number equal to approximately 2.71828. Based on this formula, what's sinh −1?

   A. $(1 - e^2)/2e$
   B. $2e - 2e^2$
   C. $(1 + e^2)/2e$
   D. $2e^2 + 2e$
   E. $4e^2 - 2e$

38. If x is a real number, then we can express the hyperbolic cosine of x as

   $$\cosh x = (e^x + e^{-x})/2$$

   Based on this formula, what's cosh −1?

   A. $(1 - e^2)/2e$
   B. $2e - 2e^2$
   C. $(1 + e^2)/2e$
   D. $2e^2 + 2e$
   E. $4e^2 - 2e$

39. In navigator's polar coordinates (NPC), we express the direction angle going

   A. counterclockwise from the positive y axis in an xy plane.
   B. clockwise from a reference axis that runs "due north" in a horizontal plane.
   C. upward from the horizon going "due north."
   D. counterclockwise from a reference axis that runs "due east" in a horizontal plane.
   E. straight out from the origin in a vertical plane.

40. In mathematician's polar coordinates (MPC), we express the direction angle going

   A. counterclockwise from the positive y axis in an xy plane.
   B. clockwise from a reference axis that runs "due north" in a horizontal plane.
   C. upward from the horizon going "due north."
   D. counterclockwise from a reference axis that runs "due east" in a horizontal plane.
   E. straight out from the origin in a vertical plane.

41. **Suppose that we see an equation where a dependent variable _r_ is a function of an independent variable _θ_ such that**

$$r = 3\theta/4$$

   If we plot this equation in MPC, we'll get a
   A. circle centered at the origin.
   B. circle not centered at the origin.
   C. straight line passing through the origin.
   D. straight line that doesn't pass through the origin.
   E. pair of spirals.

42. **Figure Exam-5 shows a system of**
   A. navigator's polar coordinates.
   B. spherical coordinates.
   C. Cartesian coordinates.
   D. cylindrical coordinates.
   E. celestial coordinates.

43. **In Fig. Exam-5, suppose that each radial division equals 10 units. In that case, we can express the coordinates of point _P_ as**
   A. $(\alpha, r) = (45°, 250\pi)$.
   B. $(\alpha, r) = (45°, 250)$.
   C. $(\alpha, r) = (250°, 45)$.
   D. $(\alpha, r) = (5\pi/4, 45)$.
   E. $(\alpha, r) = (45°, 5\pi/4)$.

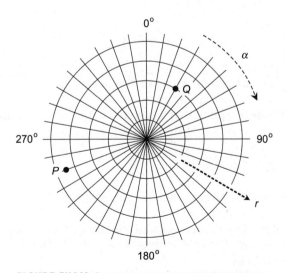

FIGURE EXAM-5 · Illustration for Final Exam Questions 42 through 44.

44. In Fig. Exam-5, suppose that each radial division equals 100 units. In that case, we can express the coordinates of point Q as

    A. $(\alpha, r) = (3, 30°)$.
    B. $(\alpha, r) = (30, 300°)$.
    C. $(\alpha, r) = (30°, 5\pi/6)$.
    D. $(\alpha, r) = (\pi/6, 30)$.
    E. $(\alpha, r) = (30°, 300)$.

45. In the MPC plane, we would express the equation of a straight line running through the origin at a right angle to the reference axis as

    A. $r = 0$.
    B. $r = \pi/2$.
    C. $\theta = \pi/2$.
    D. $r = \theta$.
    E. $r = 90°$.

46. In the MPC plane, we would express the equation of a circle with the Cartesian equation $x^2 + y^2 = 900$ as

    A. $r = 30$.
    B. $\theta = 30°$.
    C. $r = \pi/6$.
    D. $\theta = \pi/6$.
    E. $\theta = 30r$.

47. Suppose that we stand at a point on the earth's surface that lies at 45° south latitude. What's the right ascension of the point in the heavens that lies at our zenith (straight up into the sky)?

    A. 3 h
    B. −3 h
    C. 21 h
    D. 18 h
    E. We need more information to answer this question.

48. What's the latitude of a point that lies on the geographic equator?

    A. 0°
    B. +90°
    C. +180° or −180° (either one)
    D. −90°
    E. We need more information to answer this question.

49. Figure Exam-6 shows two vectors **a** and **b** in the MPC plane. Based on the information given here, we can deduce that the dot product **a** • **b**

    A. points straight at us.
    B. points straight away from us.
    C. equals 0.

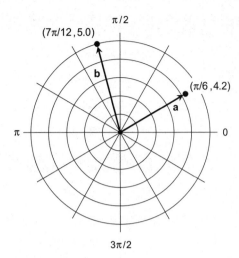

FIGURE EXAM-6 · Illustration for Final Exam
Questions 49 and 50.

    D. is a positive real number.
    E. is a negative real number.

50. **Based on the information that Fig. Exam-6 reveals, we can deduce that the cross product a × b**
    A. points straight at us.
    B. points straight away from us.
    C. equals 0.
    D. is a positive real number.
    E. is a negative real number.

51. **In Cartesian *xyz*-space, the point $(x,y,z) = (-3,4,-5)$ lies**
    A. 5 units away from the origin.
    B. 7 units away from the origin.
    C. the square root of 50 units away from the origin.
    D. the square root of 60 units away from the origin.
    E. 12 units away from the origin.

52. **In Cartesian *xyz*-space, the point $(x,y,z) = (3,-4,5)$ lies**
    A. 5 units away from the origin.
    B. 7 units away from the origin.
    C. the square root of 50 units away from the origin.
    D. the square root of 60 units away from the origin.
    E. 12 units away from the origin.

53. When we say that the cross-product operation is anticommutative, we mean that, for any two vectors **r** and **s** in three-space, one of the following equations holds true. Which one?

    A. $\mathbf{r} \times \mathbf{s} = \mathbf{s} \times \mathbf{r}$

    B. $\mathbf{r} \times \mathbf{s} = -(\mathbf{s} \times \mathbf{r})$

    C. $\mathbf{r} \bullet \mathbf{s} = \mathbf{s} \bullet \mathbf{r}$

    D. $\mathbf{r} \bullet \mathbf{s} = -(\mathbf{s} \bullet \mathbf{r})$

    E. $\mathbf{r} \times \mathbf{s} = -\mathbf{s} \times -\mathbf{r}$

54. When we say that the dot-product operation is commutative, we mean that, for any two vectors **r** and **s** in three-space, one of the following equations holds true. Which one?

    A. $\mathbf{r} \times \mathbf{s} = \mathbf{s} \times \mathbf{r}$

    B. $\mathbf{r} \times \mathbf{s} = -(\mathbf{s} \times \mathbf{r})$

    C. $\mathbf{r} \bullet \mathbf{s} = \mathbf{s} \bullet \mathbf{r}$

    D. $\mathbf{r} \bullet \mathbf{s} = -(\mathbf{s} \bullet \mathbf{r})$

    E. $\mathbf{r} \times \mathbf{s} = -\mathbf{s} \times -\mathbf{r}$

55. If we write out the expression 1.4578 e + 6 in plain decimal format, it appears as

    A. 1.4578.

    B. 14.578.

    C. 145.78.

    D. 1457.8

    E. None of the above

56. If we want to denote the plain decimal number 65,557 in standard power-of-10 notation, we can write

    A. 6.5557×10^4.

    B. 65.557×10^3.

    C. 655.57×10^2.

    D. 6555.7×10^1.

    E. Any of the above

57. Which of the following quantities exceeds 250 by five orders of magnitude?

    A. 2.50×10^{-5}

    B. 2.50×10^5

    C. 2.50×10^{-7}

    D. 2.50×10^7

    E. None of the above

58. What's the product $(1.5 \times 10^6) \times (-3.27000 \times 10^{-15})$ taking significant figures into account?

    A. -5×10^{-9}

    B. -4.9×10^{-9}

    C. -4.90×10^{-9}

    D. -4.905×10^{-9}

    E. -4.9050×10^{-9}

59. What's cos 89.99°, expressed in power-of-10 notation to four significant figures? Use a calculator and then, if necessary, convert the format.

   A. 1.745×10^{-4}
   B. 1.745×10^{-5}
   C. 1.745×10^{-6}
   D. 1.745×10^{-7}
   E. 1.745×10^{-8}

60. What's cos 89.9999°, expressed in power-of-10 notation to four significant figures? Use a calculator and then, if necessary, convert the format.

   A. 1.745×10^{-4}
   B. 1.745×10^{-5}
   C. 1.745×10^{-6}
   D. 1.745×10^{-7}
   E. 1.745×10^{-8}

61. What's cos 89.999999°, expressed in power-of-10 notation to four significant figures? Use a calculator and then, if necessary, convert the format.

   A. 1.745×10^{-4}
   B. 1.745×10^{-5}
   C. 1.745×10^{-6}
   D. 1.745×10^{-7}
   E. 1.745×10^{-8}

62. A microscopic bug crawls around on a perfectly flat parking lot. She looks over toward a tall pole. Her bug boyfriend says, "That pole is exactly 4995 millimeters tall, and its base lies 15,120 millimeters away." Based on that information, the first bug concludes that the angle between the horizon and the top of the pole, as she sees it from her vantage point at the parking-lot surface, is

   A. 17.89°.
   B. 18.28°.
   C. 19.29°.
   D. 20.20°.
   E. impossible to calculate.

63. The bug described in Question 63 looks over at another pole. Using her protractor, she measures an angle of 5.787° between the horizon and the pole's highest point. "How far away is that pole?" she asks her bug boyfriend. He says, "I don't know how far away the base of that pole is, but I've measured the distance to its top using my laser ranging apparatus, and I got a reading of 25,769 millimeters." Based on that information, the first bug concludes that the pole's height is

   A. 2824 millimeters.
   B. 2713 millimeters.
   C. 2612 millimeters.
   D. 2598 millimeters.
   E. impossible to calculate.

64. Imagine that we want to determine the distance to a star. We measure the parallax relative to the background of distant stars, which lie "infinitely far away" by comparison. We choose the times for our observations so that the earth lies directly between the sun and the star at the time of the first measurement, and a line segment connecting the sun with the star runs perpendicular to the line segment connecting the sun with the earth at the time of the second measurement (Fig. Exam-7). Suppose that the parallax thus determined equals 0.2876 seconds of arc (0.2876"). What's the distance to the star in astronomical units (AU)?

   A. 7.17×10^5
   B. 1.20×10^4
   C. 199
   D. 49
   E. We need more information to calculate it.

65. If we want to measure the distance to an object based on parallax, we must make certain that

   A. the base line is as short as possible.
   B. the base line runs perpendicular to a line connecting one observation point to the distant object.
   C. the base line runs parallel to a line connecting one observation point to the distant object.
   D. the reference objects that we use to determine the parallax are closer to us than the object whose distance we want to measure.
   E. the observed parallax is as small as possible.

66. Figure Exam-8 shows a triangle with side lengths p, q, and r. The angles opposite those sides measure θ_p, θ_q, and θ_r respectively. Which of the following equations holds true for all such triangles?

   A. $p + (\sin \theta_p) = q + (\sin \theta_q) = r + (\sin \theta_r)$
   B. $p (\sin \theta_p) = q (\sin \theta_q) = r (\sin \theta_r)$

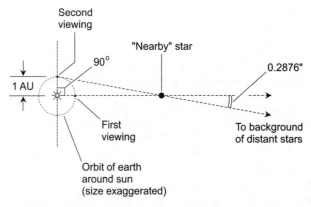

FIGURE EXAM-7 · Illustration for Final Exam Question 64.

FIGURE EXAM-8 · Illustration for Final Exam Questions 66 and 67.

C. $p/(\sin\theta_p)=q/(\sin\theta_q)=r/(\sin\theta_r)$
D. $p(\sin^2\theta_p)=q(\sin^2\theta_q)=r(\sin^2\theta_r)$
E. $p/(\sin^2\theta_p)=q/(\sin^2\theta_q)=r/(\sin^2\theta_r)$

67. In the scenario of Fig. Exam-8, suppose that $\theta_p = 90°$, and we "stretch" the triangle out by making side q longer and longer, without limit. Of course, as we carry out this operation, side p must also grow longer, and the angle θ_q must grow larger. However, no matter how long we make side q, if we want the triangle to remain "closed," the angle θ_q must always remain smaller than

A. $\pi/8$ rad.
B. $\pi/6$ rad.
C. $\pi/4$ rad.
D. $\pi/3$ rad.
E. $\pi/2$ rad.

68. Suppose that we observe two stars: Star A at a distance of 44 parsecs and Star B at a distance of 22 parsecs. From this information, we know that

A. Star B lies four times as far away as Star A.
B. Star B lies twice as far away as Star A.
C. Star A lies twice as far away as Star B.
D. Star A lies four times as far away as Star B.
E. None of the above

69. Suppose that we observe two stars: Star A at a distance of 5.5×10^6 AU and Star B at a distance of 2.2×10^7 AU. From this information we can surmise that, when we measure the parallax angles to the two stars using the radius of the earth's orbit as the base line and the "infinitely distant" stars as a background, the parallax angle of Star A will measure almost exactly

A. four times the parallax angle of Star B.
B. twice the parallax angle of Star B.
C. the same as the parallax angle of Star B.
D. half the parallax angle of Star B.
E. 1/4 of the parallax angle of Star B.

70. Figure Exam-9 is a vector diagram showing the phase relationship between the AC voltage and current in a certain electronic device at a certain frequency. The voltage and current differ by

    A. 0° of phase.
    B. 45° of phase.
    C. −45° of phase.
    D. 90° of phase.
    E. −90° of phase.

71. We can correctly say that the voltage and current in the situation of Fig. Exam-9 exist in phase

    A. quadrature.
    B. opposition.
    C. cancellation.
    D. coincidence.
    E. reactance.

72. Based on the information given in Fig. Exam-9, the represented electronic device contains

    A. pure inductive reactance.
    B. pure capacitive reactance.
    C. inductive reactance and resistance.
    D. capacitive reactance and resistance.
    E. no reactance.

73. In an electronic device, we should expect the AC current and voltage to exist in phase quadrature if and only if that device contains

    A. no reactance.
    B. pure capacitive reactance or pure inductive reactance.

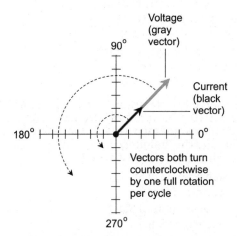

FIGURE EXAM-9 • Illustration for Final Exam Questions 70 through 72.

C. some resistance and some reactance.

D. pure resistance.

E. variable reactance.

74. **What's the period, in milliseconds (ms), of an electrical AC wave whose frequency equals 80.000 Hz? Remember that one millisecond equals precisely 0.001 second.**

A. 1.2500 ms

B. 0.80000 ms

C. 80.000 ms

D. 12.500 ms

E. 640.00 ms

75. **For the wave described in Question 74, how much time represents exactly $\pi/3$ rad of phase?**

A. 2.0833 ms

B. 0.20000 ms

C. 13.333 ms

D. 320.00 ms

E. 6.2500 ms

76. **In an AC sine wave, 135° of phase corresponds to**

A. 1/8 of a cycle.

B. 1/4 of a cycle.

C. 1/2 of a cycle.

D. 3/8 of a cycle.

E. 3/4 of a cycle.

77. **If the current lags the voltage by 1/9 of a cycle in a certain component at a specified, constant AC frequency, then the phase angle equals**

A. 60°.

B. 45°.

C. 40°.

D. 33°.

E. 30°.

78. **Imagine that a light ray travels through a clear solid substance whose index of refraction equals r. The ray encounters a flat boundary between the original substance and another clear solid substance whose index of refraction equals s. As the angle of incidence (with respect to a normal line) starts out at 0° and gradually increases, we can define the critical angle θ_c, at which total internal reflection first occurs, if and only if**

A. $r < s$.

B. $r \leq s$.

C. $r = s$.

D. $r \geq s$.

E. $r > s$.

79. In a situation of the sort described in Question 78 where total internal reflection can actually take place, we can calculate the critical angle θ_c as

$$\theta_c = \text{Arcsin}\,(s/r)$$

where r and s represent the indices of refraction. Based on this formula and definition, what's the critical angle for a ray of light traveling through a sample of calcite (refractive index 1.66) and encountering a flat barrier with fresh water ice (refractive index 1.31)?

A. 49.5°
B. 52.1°
C. 45.0°
D. 40.5°
E. 37.9°

80. Suppose that, in the scenario of Question 79, the fresh water ice melts into liquid, changing its refractive index to 1.33. The critical angle will

A. vanish altogether.
B. decrease slightly.
C. decrease a lot.
D. increase slightly.
E. increase a lot.

81. Figure Exam-10 shows how a concave mirror forms a real image at a focal point. Suppose that the initial distance r_o between the mirror and the object equals 1.2 times the mirror's focal length f. If we increase r_o to $12f$, what will happen to the distance r_i between the mirror and the object's real image?

A. We can't define it because the image will vanish.
B. It will increase slightly.
C. It will increase a lot.
D. It will decrease slightly.
E. It will decrease a lot.

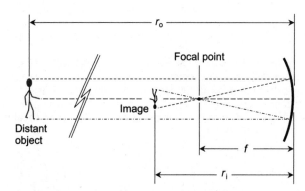

FIGURE EXAM-10 · Illustration for Final Exam Questions 81 and 82.

82. In the situation of Question 81 and Fig. Exam-10, suppose that the initial distance r_o between the mirror and the object equals 1.2 times the mirror's focal length f. If we halve r_o to $0.6f$, what will happen to the distance r_i between the mirror and the object's image?

   A. We can't define it because the image will vanish.
   B. It will increase slightly.
   C. It will increase a lot.
   D. It will decrease slightly.
   E. It will decrease a lot.

83. Figure Exam-11 shows a ray of light passing through a flat boundary from air with a refractive index of 1.00 to a sample of ruby with a refractive index of 1.76. According to Snell's law, the sine of the lower angle ϕ divided by the sine of the upper angle of 60° equals

   A. Arcsin (1.00/1.76).
   B. 1.00/1.76.
   C. 1.76.
   D. Arctan 1.76.
   E. (1.00/1.76)².

84. In the situation of Fig. Exam-11, angle ϕ measures

   A. 16.5°
   B. 21.5°.
   C. 29.5°.
   D. 33.0°.
   E. 38.8°.

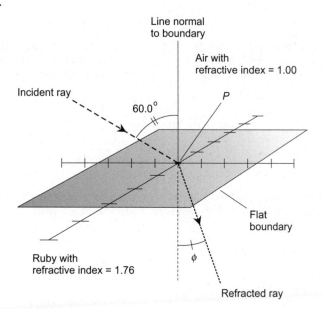

FIGURE EXAM-11 · Illustration for Final Exam Questions 83 and 84.

85. When parallel rays of light strike a matte surface, the reflected rays
    A. emerge parallel to each other.
    B. converge toward a point.
    C. diverge.
    D. emerge in a random way.
    E. emerge perpendicular to the surface at every point.

86. Based on the definitions of meridians (longitude arcs) and parallels (latitude circles) given in this book, we can deduce that on the earth's surface, the meridian for 90° east longitude intersects the parallel for 30° south latitude at an angle of
    A. $\pi/4$ rad.
    B. $\pi/2$ rad.
    C. π rad.
    D. 2π rad.
    E. undefinable extent.

87. Suppose that we extend a geodesic arc on the earth's surface to form a complete great circle that passes through both geographic poles. This circle defines
    A. one parallel.
    B. two parallels.
    C. one meridian.
    D. two meridians.
    E. the geographic equator.

88. Figure Exam-12 shows a triangle on the surface of the earth, with sides running along the prime meridian, the 120° east longitude meridian, and the equator. What's the sum of the measures of the interior angles of this triangle?
    A. 360°
    B. 330°
    C. 300°
    D. 270°
    E. 240°

89. In the scenario of Fig. Exam-12, what's the tangent of the angle connecting the side that runs along the 120° east longitude meridian with the side running along the equator?
    A. −1.732
    B. −1.000
    C. −0.667
    D. 0.500
    E. It's undefined.

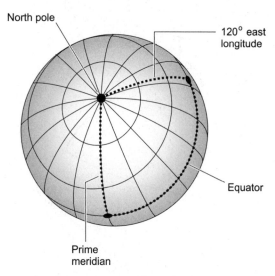

North pole

120° east
longitude

Equator

Prime
meridian

FIGURE EXAM-12 · Illustration for Final Exam
Questions 88 and 89.

90. Suppose that we draw a spherical quadrilateral (four-sided geometric figure) on the surface of the earth, so that all four of the vertices lie along a geodesic that runs through both of the earth's poles. In that case, the perimeter of the spherical quadrilateral equals

  A. the earth's circumference.
  B. $2/\pi$ times the earth's circumference.
  C. half the earth's circumference.
  D. $1/\pi$ times the earth's circumference.
  E. $\pi/4$ times the earth's circumference.

91. If we add up the sines of the four interior angles of the spherical quadrilateral described in Question 90, we'll get

  A. zero.
  B. 0.500.
  C. 0.866.
  D. 1.000.
  E. 1.732.

92. In the scenario of Questions 90 and 91, the exterior area of the spherical quadrilateral equals

  A. infinity (or, if you prefer, an undefined value).
  B. four times the earth's surface area.
  C. 1/4 of the earth's surface area.
  D. the spherical quadrilateral's interior area.
  E. zero.

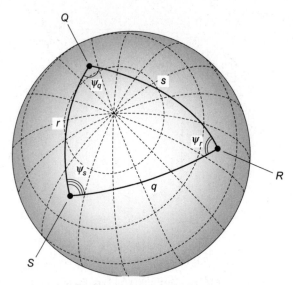

FIGURE EXAM-13 · Illustration for Final Exam Question 93.

93. Consider a spherical triangle with vertices Q, R, and S. Call the angular extents of the sides opposite each vertex point, expressed in radians, by the names q, r, and s respectively (Fig. Exam-13). Call the interior spherical angles ψ_q, ψ_r, and ψ_s as shown. In this situation, we can have absolute confidence that

A. $\sin \psi_q = \sin \psi_r = \sin \psi_s$.
B. $\sin q = \sin r = \sin s$.
C. $(\sin q)/(\sin \psi_q) = (\sin r)/(\sin \psi_r) = (s \in s)/(\sin \psi_s)$.
D. $(\sin q)(\sin \psi_q) = (\sin r)(\sin \psi_r) = (\sin s)(\sin \psi_s)$.
E. $(\sin q)+(\sin \psi_q) = (\sin r)+(\sin \psi_r) = (\sin s)+(\sin \psi_s)$.

94. Consider an infinite series F in which the nth term equals the reciprocal of 4n, as follows:

$$F = 1/4 + 1/8 + 1/12 + 1/16 + 1/20 + 1/24 + \cdots$$

As we go out in this series indefinitely, the term value

A. grows without bound.
B. approaches $4n^2$.
C. approaches $(4n+1)/4n$.
D. approaches $n^4/n!$.
E. approaches 0.

95. Consider an infinite series G in which the nth term equals the reciprocal of n^4, as follows:

$$F = 1/ + 1/16 + 1/81 + 1/256 + 1/625 + 1/1296 + \cdots$$

As we go out in this series indefinitely, the term value

A. grows without bound.
B. approaches $4n^2$.
C. approaches $(4n+1)/4n$.
D. approaches $n^4/n!$.
E. approaches 0.

96. When the terms of an infinite series "blow down" (skyrocket negatively) as we keep adding up partial sums, we say that the series

A. contradicts the limit.
B. diverges.
C. crosses an asymptote.
D. converges to negative infinity.
E. limits itself negatively.

97. As we go out indefinitely in the series expansion for cos 120°, the term value

A. increases negatively without limit.
B. approaches 0.
C. approaches $-1/2$.
D. approaches $-1/3$.
E. approaches $-1/\pi$.

98. As we go out indefinitely in the series expansion for sin $5\pi/6$ rad, the term value

A. increases positively without limit.
B. approaches 0.
C. approaches $1/2$.
D. approaches $1/3$.
E. approaches $1/\pi$.

99. In a geometric sequence, the subtractive difference between any given term and the next one always

A. remains constant as we move along.
B. keeps getting smaller as we move along.
C. keeps getting larger as we move along.
D. alternates between positive and negative values as we move along.
E. None of the above

100. In the range of angles θ in radians where $0 \le \theta \le \pi/2$, the infinite series expansion for the cosine function

A. diverges if and only if $\theta = 0$.
B. diverges if and only if $\theta = \pi/4$.
C. diverges if and only if $\theta = \pi/3$.
D. diverges if and only if $\theta = \pi/2$.
E. never diverges.

Answers to Quizzes, Tests, and Final Exam

Chapter 1	Chapter 3	Chapter 5	Chapter 7
1. B	1. B	1. C	1. D
2. A	2. C	2. A	2. B
3. B	3. A	3. A	3. B
4. D	4. D	4. C	4. C
5. C	5. D	5. D	5. D
6. A	6. A	6. C	6. A
7. A	7. D	7. B	7. A
8. D	8. A	8. B	8. C
9. C	9. A	9. B	9. C
10. C	10. B	10. D	10. D

Chapter 2	Chapter 4	Chapter 6	Test: Part I
1. D	1. D	1. C	1. B
2. B	2. A	2. D	2. C
3. D	3. D	3. C	3. B
4. A	4. A	4. B	4. C
5. C	5. C	5. A	5. D
6. C	6. D	6. A	6. E
7. B	7. A	7. B	7. A
8. C	8. C	8. D	8. B
9. A	9. C	9. D	9. D
10. A	10. A	10. C	10. A

11. B	49. C	**Chapter 11**	2. C
12. D	50. B	1. C	3. B
13. C		2. B	4. A
14. A	**Chapter 8**	3. C	5. A
15. A	1. D	4. A	6. C
16. C	2. A	5. D	7. E
17. B	3. B	6. D	8. E
18. A	4. C	7. C	9. A
19. A	5. D	8. B	10. C
20. D	6. A	9. B	11. C
21. A	7. C	10. D	12. B
22. B	8. C		13. A
23. E	9. D	**Chapter 12**	14. B
24. E	10. A	1. C	15. D
25. A		2. C	16. D
26. D	**Chapter 9**	3. A	17. B
27. C	1. C	4. C	18. E
28. A	2. B	5. D	19. D
29. D	3. A	6. B	20. E
30. A	4. A	7. B	21. C
31. E	5. B	8. A	22. E
32. B	6. C	9. A	23. A
33. C	7. D	10. B	24. B
34. A	8. D		25. A
35. D	9. B	**Chapter 13**	26. B
36. D	10. B	1. A	27. A
37. E		2. C	28. D
38. D	**Chapter 10**	3. D	29. B
39. C	1. C	4. D	30. D
40. A	2. A	5. A	31. C
41. B	3. A	6. B	32. B
42. A	4. D	7. B	33. D
43. C	5. B	8. D	34. B
44. D	6. B	9. D	35. A
45. E	7. C	10. B	36. C
46. B	8. A		37. D
47. E	9. C	**Test: Part II**	38. E
48. B	10. B	1. C	39. C

40. E
41. D
42. D
43. C
44. E
45. A
46. B
47. B
48. C
49. B
50. A

Final Exam
1. A
2. C
3. B
4. D
5. A
6. A
7. D
8. E
9. C
10. C
11. A
12. E
13. C
14. B
15. A
16. A
17. B
18. D
19. E
20. B
21. A
22. E
23. C
24. D
25. C
26. B

27. A
28. D
29. B
30. C
31. A
32. B
33. B
34. A
35. D
36. C
37. A
38. C
39. B
40. D
41. E
42. A
43. C
44. E
45. C
46. A
47. E
48. A
49. D
50. A
51. C
52. C
53. B
54. C
55. E
56. A
57. D
58. B
59. A
60. C
61. E
62. B
63. D
64. A

65. B
66. C
67. E
68. C
69. A
70. A
71. D
72. E
73. B
74. D
75. A
76. D
77. C
78. E
79. B
80. D
81. E
82. A
83. B
84. C
85. D
86. B
87. D
88. C
89. E
90. A
91. A
92. D
93. C
94. E
95. E
96. B
97. C
98. C
99. E
100. E

Circular and Hyperbolic Identities

Trigonometric identities can help you solve complicated or tricky problems in the real world. The following basic facts can help you "crack" difficult expressions by getting them into forms that lend themselves to calculation.

Identities for Circular Functions

Unless otherwise specified, the following formulas all apply to angles θ and ϕ in the standard range:

$$0 \le \theta < 2\pi$$
$$0° \le \theta < 360°$$
$$0 \le \phi < 2\pi$$
$$0° \le \phi < 360°$$

You can convert angles outside the standard range to values within the standard range by adding or subtracting the appropriate multiple of $360°$ (2π).

Pythagorean Theorem for Sine and Cosine

The sum of the squares of the sine and cosine of an angle θ equals 1 for all possible values of θ, as follows:

$$\sin^2 \theta + \cos^2 \theta = 1$$

Pythagorean Theorem for Secant and Tangent

The difference between the squares of the secant and tangent of θ equals 1 or −1 for all values of θ where the secant and the tangent are defined, as follows:

$$\sec^2 \theta - \tan^2 \theta = 1$$

and

$$\tan^2 \theta - \sec^2 \theta = -1$$

Pythagorean Theorem for Cosecant and Cotangent

The difference between the squares of the cosecant and cotangent θ equals 1 or −1 for all values of θ where the cosecant and the cotangent are defined, as follows:

$$\csc^2 \theta - \cot^2 \theta = 1$$

and

$$\cot^2 \theta - \csc^2 \theta = -1$$

Sine of Negative Angle

The sine of the negative of an angle (an angle measured in the direction opposite to the normal direction) equals the negative (additive inverse) of the sine of the angle. The following formula holds:

$$\sin -\theta = -\sin \theta$$

Cosine of Negative Angle

The cosine of the negative of an angle equals the cosine of the angle. The following formula holds:

$$\cos -\theta = \cos \theta$$

Tangent of Negative Angle

The tangent of the negative of an angle equals the negative (additive inverse) of the tangent of the angle. The following formula applies for all angles except $\theta = 90°\ (\pi/2\ \text{rad})$ and $\theta = 270°\ (3\pi/2\ \text{rad})$:

$$\tan -\theta = -\tan \theta$$

Cosecant of Negative Angle

The cosecant of the negative of an angle equals the negative (additive inverse) of the cosecant of the angle. The following formula applies for all angles except $\theta = 0°$ (0 rad) and $\theta = 180°$ (π rad):

$$\csc -\theta = -\csc \theta$$

Secant of Negative Angle

The secant of the negative of an angle equals the secant of the angle. The following formula applies for all angles except $\theta = 90°$ ($\pi/2$ rad) and $\theta = 270°$ ($3\pi/2$ rad):

$$\sec -\theta = \sec \theta$$

Cotangent of Negative Angle

The cotangent of the negative of an angle equals the negative (additive inverse) of the cotangent of the angle. The following formula applies for all angles except $\theta = 0°$ (0 rad) and $\theta = 180°$ (π rad):

$$\cot -\theta = -\cot \theta$$

Sine of Double Angle

The sine of twice any given angle equals twice the sine of the original angle times the cosine of the original angle:

$$\sin 2\theta = 2 \sin \theta \cos \theta$$

Cosine of Double Angle

The cosine of twice any given angle can be found according to either of the following:

$$\cos 2\theta = 1 - (2 \sin^2 \theta)$$

or

$$\cos 2\theta = (2 \cos^2 \theta) - 1$$

Sine of Angular Sum

The sine of the sum of two angles θ and ϕ can be found using this formula:

$$\sin (\theta + \phi) = (\sin \theta)(\cos \phi) + (\cos \theta)(\sin \phi)$$

Cosine of Angular Sum

The cosine of the sum of two angles θ and ϕ can be found using this formula:

$$\cos(\theta + \phi) = (\cos\theta)(\cos\phi) - (\sin\theta)(\sin\phi)$$

Sine of Angular Difference

The sine of the difference between two angles θ and ϕ can be found using this formula:

$$\sin(\theta - \phi) = (\sin\theta)(\cos\phi) - (\cos\theta)(\sin\phi)$$

Cosine of Angular Difference

The cosine of the difference between two angles θ and ϕ can be found using this formula:

$$\cos(\theta - \phi) = (\cos\theta)(\cos\phi) + (\sin\theta)(\sin\phi)$$

Identities for Hyperbolic Functions

The following formulas all apply to real numbers x and y, with certain restrictions in special cases where indicated.

Pythagorean Theorem for Sinh and Cosh

The difference between the squares of the hyperbolic sine and hyperbolic cosine of a variable is always equal to either 1 or −1. The following formulas hold for all real numbers x:

$$\sinh^2 x - \cosh^2 x = -1$$

and

$$\cosh^2 x - \sinh^2 x = 1$$

Pythagorean Theorem for Csch and Coth

The difference between the squares of the hyperbolic cotangent and hyperbolic cosecant of a variable is always equal to either 1 or −1. The following formulas hold for all real numbers x except 0:

$$\operatorname{csch}^2 x - \coth^2 x = -1$$

and

$$\coth^2 x - \operatorname{csch}^2 x = 1$$

Pythagorean Theorem for Sech and Tanh

The sum of the squares of the hyperbolic secant and hyperbolic tangent of a variable is always equal to 1. The following formula holds for all real numbers x:

$$\text{sech}^2 x + \tanh^2 x = 1$$

Hyperbolic Sine of Negative Variable

The hyperbolic sine of the negative of a variable equals the negative of the hyperbolic sine of the variable. The following formula holds for all real numbers x:

$$\sinh -x = -\sinh x$$

Hyperbolic Cosine of Negative Variable

The hyperbolic cosine of the negative of a variable equals the hyperbolic cosine of the variable. The following formula holds for all real numbers x:

$$\cosh -x = \cosh x$$

Hyperbolic Tangent of Negative Variable

The hyperbolic tangent of the negative of a variable equals the negative of the hyperbolic tangent of the variable. The following formula holds for all real numbers x:

$$\tanh -x = -\tanh x$$

Hyperbolic Cosecant of Negative Variable

The hyperbolic cosecant of the negative of a variable equals the negative of the hyperbolic cosecant of the variable. The following formula holds for all real numbers x except 0:

$$\text{csch} -x = -\text{csch}\, x$$

Hyperbolic Secant of Negative Variable

The hyperbolic secant of the negative of a variable equals the hyperbolic secant of the variable. The following formula holds for all real numbers x:

$$\text{sech} -x = \text{sech}\, x$$

Hyperbolic Cotangent of Negative Variable

The hyperbolic cotangent of the negative of a variable equals the negative of the hyperbolic cotangent of the variable. The following formula holds for all real numbers x except 0:

$$\coth -x = -\coth x$$

Hyperbolic Sine of Double Value

The hyperbolic sine of twice any given variable equals twice the hyperbolic sine of the original variable times the hyperbolic cosine of the original variable. The following formula holds for all real numbers x:

$$\sinh 2x = 2 \sinh x \cosh x$$

Hyperbolic Cosine of Double Value

The hyperbolic cosine of twice any given variable can be found according to any of the following three formulas for all real numbers x:

$$\cosh 2x = \cosh^2 x + \sinh^2 x$$
$$\cosh 2x = 1 + 2 \sinh^2 x$$
$$\cosh 2x = 2 \cosh^2 x - 1$$

Hyperbolic Sine of Sum

The hyperbolic sine of the sum of two variables x and y can be found according to the following formula for all real numbers x and y:

$$\sinh (x + y) = \sinh x \cosh y + \cosh x \sinh y$$

Hyperbolic Cosine of Sum

The hyperbolic cosine of the sum of two variables x and y can be found according to the following formula for all real numbers x and y:

$$\cosh (x + y) = \cosh x \cosh y + \sinh x \sinh y$$

Hyperbolic Sine of Difference

The hyperbolic sine of the difference between two variables x and y can be found according to the following formula for all real numbers x and y:

$$\sinh(x - y) = \sinh x \cosh y - \cosh x \sinh y$$

Hyperbolic Cosine of Difference

The hyperbolic cosine of the difference between two variables x and y can be found according to the following formula for all real numbers x and y:

$$\cosh(x - y) = \cosh x \cosh y - \sinh x \sinh y$$

Suggested Additional Reading

Downing, Douglas, *Trigonometry the Easy Way*. Hauppauge, NY: Barron's Educational Series, 2001.

Gibilisco, Stan, *Algebra Know-It-All*. New York, NY: McGraw-Hill, 2008.

Gibilisco, Stan, *Geometry Demystified*, 2nd ed. New York, NY: McGraw-Hill, 2011.

Gibilisco, Stan, *Mastering Technical Mathematics*, 3rd ed. New York, NY: McGraw-Hill, 2007.

Gibilisco, Stan, *Math Proofs Demystified*. New York, NY: McGraw-Hill, 2005.

Gibilisco, Stan, *Pre-Calculus Know-It-All*. New York, NY: McGraw-Hill, 2010.

Gibilisco, Stan, *Technical Math Demystified*. New York, NY: McGraw-Hill, 2006.

Huettenmueller, Rhonda, *Algebra Demystified*, 2nd ed. New York, NY: McGraw-Hill, 2011.

Kay, David A., *CliffsQuickReview Trigonometry*. Hoboken, NJ: Wiley Publishing, Inc., 2001.

Learning Express Editors, *Trigonometry Success in 20 Minutes a Day*. Devens, MA: Learning Express LLC, 2007.

Moyer, Robert, and Ayres, Frank, *Schaum's Outline of Trigonometry*, 4th ed. New York, NY: McGraw-Hill, 2008.

Prindle, Anthony and Katie, *Math the Easy Way*, 2nd ed. Hauppauge, NY: Barron's Educational Series, 2009.

Ross, Debra Anne, *Master Math: Trigonometry*. Florence, KY: Course Technology, a part of Cengage Learning, 2009.

Sterling, Mary Jane, *Trigonometry for Dummies*. Hoboken, NJ: Wiley Publishing, Inc., 2005.

Sterling, Mary Jane, *Trigonometry Workbook for Dummies*. Hoboken, NJ: Wiley Publishing, Inc., 2005.

Index

A

absolute error, 210–211
acute angle, 5
acute triangle, 13–14
addition in scientific notation, 204–205
adjacent sides, 21
alpha, as angle name, 4
alternate interior angles, 226
alternating current, 248–252
alternative scientific notation, 197–198
amplitude, 56, 251
angle
  acute, 5
  critical, 281–286
  interior, 12
  naming of, 4
  notation for, 7–8
  obtuse, 5
  optical, 270
  reflex, 5

angle (*Cont.*):
  right, 3, 5
  straight, 5
angle bisection principle, 8
angle of convergence, 274
angle of incidence, 270–272, 281–287
angle of reflection, 270–272
angular addition, 10–11
angular difference
  cosine of, 398
  sine of, 398
angular frequency, 251–252
angular resolution, 225
angular sides, 308–310
angular subtraction, 10–11
angular sum
  cosine of, 398
  sine of, 397
anticommutative property, 172
antipodes, 150
arc minute, definition of, 6
arc of a great circle, definition of, 298
arc second, definition of, 6

Arccosecant function
definition of, 91
graph of, 95–96
Arccosine function
definition of, 90
graph of, 94–95
Arccotangent function
definition of, 91
graph of, 96
Arcsecant function
definition of, 91
graph of, 95–96
Arcsine function
definition of, 90
graph of, 93–94
Arctangent function
definition of, 90
graph of, 95
use of, in coordinate transformation,
133–134
argument of a function, 52
arithmetic mean, 43
arithmetic operations, precedence of,
211–213
arithmetic progression, 327
arithmetic sequence, 326–330
arithmetic series, 327–330
astronomical unit
definition of, 229
use of, 231–232
asymptote, 105
average, 43
azimuth in spherical coordinates, 156

B

base line for distance measurement,
224–225

beta, as angle name, 4
bijection, 81–83
bijective mapping, 81–83
bijective relation, 85–86
bilateral symmetry, 105

C

Cantor, Georg, 61
capacitive reactance, 262–265
capacitor, 262
Cartesian coordinates
in two dimensions, 27–49
Cartesian distance between two
points, 38–42
Cartesian plane
definition of, 28
quadrants in, 31–32
variables in, 83–84
vectors in, 159–163
vs. rectangular coordinates, 32–33
Cartesian three-space
definition of, 154
vectors in, 168–176
Cartesian to polar coordinate
transformation, 133–138, 145–146
celestial equator, 152
celestial latitude, 151, 156
celestial longitude, 151, 156
celestial poles, 152
circle centered at origin, 126
circle in polar coordinates, 125–126,
139–141
circular functions
identities for, 395–398
inverses of, 88–99
primary, 54–62
secondary, 62–71

closed interval, 30

codomain, 76–79

collinear points, 11

color dispersion, 287–293

commutative principle, 161

congruent triangles, 13

constant-angle graph, 125

constant-radius graph, 125–126

concave mirror

  focal length of, 275

  focal point of, 275

convergence of sequence or series,
    333–334

convex mirror, 274–275

coordinate transformation

  Cartesian to polar, 133–138,
     145–146

  polar to Cartesian, 131–133, 144

  polar to polar, 143–145

coordinates

  Cartesian, 27–49

  celestial, 151–153

  cylindrical, 154–155

  polar, 121–148

  spherical, 155–157

  terrestrial, 150–151

cosecant function, 62–64

cosecant of negative angle, 397

cosine function

  definition of, 56–59

  series for, 341–344

cosine of angular difference, 398

cosine of angular sum, 398

cosine of double angle, 397

cosine of negative angle, 396

cosine wave, 58–59

cotangent function, 66–67

cotangent of negative angle, 397

critical angle, 281–286

cross product of two vectors,
    171–176, 214–215

cycle of a wave, 248–249

cycles per second, 249

cylindrical coordinates, 154–155,
    158–159

D

declination, 152–153, 156

degree

  angular, definition of, 5–6

  phase, 249–250

delta, meaning "the difference in," 40

dependent variable, 28, 83–84

direction angle, 52–54

direction in polar coordinates,
    122–123

direction of vector, 159–161,
    163–164, 169

directly congruent triangles, 13

directly similar triangles, 12

dish antenna, 233

dispersion, color, 287–293

distance addition, 9–10

distance between two points

  Cartesian, 38–42

  over the surface of a sphere, 298

  polar, 138–139

distance measurement

  base line for, 224–225

  interstellar, 229–232

  terrestrial, 224–229

distance of point from origin in
    Cartesian plane, 34–38

distance subtraction, 9–10

division in scientific notation, 206–207

domain of mapping, 76–79
dot product of two vectors, 162–163,
    166–167, 171, 213–214
double angle
  cosine of, 397
  sine of, 397
  tangent of, 397
double value
  hyperbolic cosine of, 400
  hyperbolic sine of, 400

E

electric field, 262
electrical resistance, 258–259
electromagnetic field, 248
elevation in spherical coordinates,
    156
equator, celestial, 152
equatorial axis, 150
equilateral spherical triangle, 313–316
equilateral triangle, 15–16
error determination, 210–211
essential domain, 77
Euclidean surface, 3
Euclidean universe, 11
excluded value, 30
expanding spherical triangle, 310–312
exponential function, 106
exponentiation in scientific notation,
    208

F

factorial, 336–337
finite sequence, definition of, 326
flat mirror, 273–274
focal length of concave mirror, 275

focal point of concave mirror, 275
"fox hunt," 233–235
frequency
  angular, 251–252
  capacitive reactance and, 262–263
  inductive reactance and, 259–260
  sine-wave, 56, 249–252
function
  definition of, 30, 86
  examples of, 86–88
  inverse, 88–89, 92–93
  vertical line test for, 97
functions, trigonometric, definitions
    of, 17–18

G

gamma, as angle name, 4
geodesic, definition of, 298
geodesic, minor, definition of,
    305–306
geodesic arc, definition of, 298
geographic north, 142
geomagnetic north, 142
geometric sequence, 332–334
geometric series, 332–334
geometrical optics, 269–296
gigacycle, definition of, 249
gigahertz, definition of, 249
global navigation, 316–321
great circle, definition of, 298
Greenwich meridian, 150–151,
    299–300

H

half-open interval, 30
hertz, definition of, 249

hour of right ascension, definition of, 152–153

hyperbola, unit, definition of, 104–105

hyperbolic angle, 104–105

hyperbolic Arccosecant
definition of, 112–114
graph of, 116

hyperbolic Arccosine
definition of, 112–114
graph of, 115

hyperbolic Arccotangent
definition of, 112–114
graph of, 117

hyperbolic Arcsecant
definition of, 112–114
graph of, 116

hyperbolic Arcsine
definition of, 112–114
graph of, 114

hyperbolic Arctangent
definition of, 112–114
graph of, 115

hyperbolic cosecant
definition of, 107
graph of, 109–110

hyperbolic cosecant of negative variable, 399

hyperbolic cosine
exponential definition of, 106
geometric definition of, 104–106
graph of, 107–108

hyperbolic cosine of difference, 401

hyperbolic cosine of double value, 400

hyperbolic cosine of negative variable, 399

hyperbolic cosine of sum, 400

hyperbolic cotangent
definition of, 107
graph of, 110–111

hyperbolic cotangent of negative variable, 400

hyperbolic functions, identities for, 398–401

hyperbolic secant
definition of, 107
graph of, 110

hyperbolic secant of negative variable, 399

hyperbolic sine
exponential definition of, 106
geometric definition of, 104–106
graph of, 107–108

hyperbolic sine of difference, 401

hyperbolic sine of double value, 400

hyperbolic sine of negative variable, 399

hyperbolic sine of sum, 400

hyperbolic tangent
definition of, 107
graph of, 108–109

hyperbolic tangent of negative variable, 399

I

identities for circular functions, 395–398

identities for hyperbolic functions, 398–401

incidence, angle of, 270–272, 281–287

included value, 30

independent variable, 28, 83–84

index of refraction, 279–280

inductive reactance, 258–261

inductor, 259
infinity, 61
inflection point, 61
injection, 80
injective mapping, 80
injective relation, 84
instantaneous amplitude, 251
interior angle, 12
interstellar distance measurement,
    229–232
interval
    closed, 30
    half-open, 30
    notation for, 29–30
    open, 30
inverse function, 88–89, 92–93
inverse relation, 88
inverses of circular functions, 88–99
irrational number, 4, 106
isosceles triangle, 14–15

K

kilocycle, definition of, 249
kilohertz, definition of, 249
kilometers per arc minute of
    latitude, 302
kilometers per arc minute of
    longitude, 303
kilometers per arc second of
    latitude, 302
kilometers per arc second of
    longitude, 303
kilometers per degree of
    latitude, 302
kilometers per degree of
    longitude, 303
kiloparsec, definition of, 230

L

lagging phase, 254–256
latitude
    celestial, 151, 156
    distance per unit of, 302–305
    terrestrial, 150–151, 239–243,
        299–305
law of cosines
    on Euclidean plane, 236–239
    on sphere, 312–313
law of reflection, 270–272
law of sines
    on Euclidean plane, 235–236, 242
    on sphere, 312
leading phase, 254–256
lemma, definition of, 44
light year, definition of, 229
limits of sequences and series, 335–337
lines
    normal, 4, 270
    orthogonal, 4
    perpendicular, 4
locally tangent plane, 270
logarithm, natural, 106
longitude
    celestial, 151, 156
    distance per unit of, 302–305
    terrestrial, 150–151, 239–243,
        299–305
loop antenna, 234

M

magnetic field, 248
magnitude of vector, 159–161,
    163–164, 169, 173
many-to-one relation, 87

mapping
  bijective, 81–83
  definition of, 76–79
  injective, 80
  one-to-one, 80
  onto, 80–81
  surjective, 80–81
  types of, 79–83
mathematician's cylindrical coordinates,
    154–155
maximal domain, 76–79
megacycle, definition of, 249
megahertz, definition of, 249
megaparsec, definition of, 230
meridian, prime, 151
meridians on earth's surface, 300–301
midpoint between two points in
    Cartesian plane, 43–46
minor geodesic, definition of, 305–306
minute of arc, definition of, 6
minute of right ascension, definition
    of, 152–153
mirror
  concave, 275–279
  convex, 274–275
  flat, 273–274
  paraboloidal, 275–276
  spherical, 275–276
mirror-image spirals, 126–130
multiplication in scientific notation,
    206
multiplication symbol in scientific
    notation, 198

N

nadir in spherical coordinates, 156–157
natural logarithm, 106

navigation, global, 316–321
navigator's cylindrical coordinates,
    154, 158–159
navigator's polar coordinates,
    142–146
negative angle
  cosecant of, 397
  cosine of, 396
  cotangent of, 397
  secant of, 397
  sine of, 396
  tangent of, 396
negative direction angle, 54
negative variable
  hyperbolic cosecant of, 399
  hyperbolic cosine of, 399
  hyperbolic cotangent of, 400
  hyperbolic secant of, 399
  hyperbolic sine of, 399
  hyperbolic tangent of, 399
normal lines, 4, 270
north celestial pole, 152

O

obtuse angle, 5
obtuse triangle, 13–14
"offbeat" direction angle, 52–54
ohm, 259
one-to-many relation, 87
one-to-one correspondence, 82
one-to-one mapping, 80
open interval, 30
optics, geometrical, 269–296
ordered pair, 28–29, 77–79
ordered triple, 153
orders of magnitude, 200–203
origin in Cartesian coordinates, 28

origin in rectangular three-space, 153
orthogonal lines, 4

P

parabola, 89–90
paraboloidal mirror, 275–276
parallax, 224–229
parallels on earth's surface, 300–301
parsec
  definition of, 229
  use of, 229–232
partial sums, definition of, 329
peak positive amplitude, 251
period of alternating-current wave:
  248
perpendicular bisector principle,
  8–9
perpendicular lines, 4
perpendicular principle, 8
phase
  degrees of, 250
  lagging, 254–256
  leading, 254–256
  radians of, 250–251
  vectors showing, 256–257
phase angle, 252–258, 260–261,
  264–265
phase coincidence, 252–253
phase difference, 252–258
phase opposition, 254
phi, as angle name, 4
plain-text exponents in scientific
  notation, 199
plane, locally tangent, 270
point matching, 76–77
point of inflection, 61
points, collinear, 11

polar axis, 150
polar coordinates
  circle in, 125–126, 139–141
  constant-angle graph in, 125
  constant-radius graph in, 125–126
  direction in, 122–123
  functions in, 123–124
  mirror-image spirals in, 126–130
  navigator's, 142–146
  radius in, 122
  relations in, 123–124
  spirals in, 126–130
  straight line in, 125, 129–130,
   139–141
  variables in, 122–124
  vectors in, 213–215
polar distance between two points,
  138–139
polar to Cartesian coordinate
  transformation, 131–133, 144
polar to polar coordinate
  transformation, 143–145
positive or negative infinity, 61
power-of-10 notation. *See* scientific
  notation
precedence of arithmetic operations,
  211–213
prefix multipliers in scientific notation,
  202–203
principal branch, 94–98
principal values, 94–99
primary circular functions, 54–62
prime meridian, 151, 299–300
prism, color dispersion in, 287–293
proper subset, 82
proportional error, 210–211
proportional error percentage,
  210–211

Pythagorean theorem
  for cosecant and cotangent, 69–70, 396
  for hyperbolic cosecant and cotangent, 398
  for hyperbolic secant and tangent, 399
  for hyperbolic sine and cosine, 398
  for right triangles, 18–21, 35
  for secant and tangent, 67–68, 396
  for sine and cosine, 23–24, 67–71, 396
  modified, 237
Pythagoras of Samos, 18

Q

quadrants in Cartesian plane, 31–32

R

radar, 233–234
radial distance of point from origin in Cartesian plane, 34–38
radian
  definition of, 4–5
  phase, 249–250
radio detection and ranging, 233–234
radio direction finding, 233–234
radiolocation, 235–236
radius in polar coordinates, 122
radius of curvature, 274–275
"rainbow" spectra, 288–293
range in spherical coordinates, 156
range of mapping, 76–79
rational number, 79
RC phase angle, 264–265

reactance
  capacitive, 262–265
  inductive, 258–261
real image with concave mirror, 276–279
real number, 79
rectangular coordinates vs. Cartesian plane, 32–33
rectangular three-space, 153–154
reference axis in cylindrical coordinates, 154
reflection
  angle of, 270–272
  law of, 270–272
  mirrors and, 270–279
  total internal, 281–282
reflex angle, 5
refraction
  angle of, 281–287
  index of, 279–280
  Snell's law for, 283–284
refractive index, 279–280
relation
  bijective, 85–86
  definition of, 30
  examples of, 83–86
  injective, 84
  many-to-one, 87
  one-to-many, 87
  surjective, 84–85
resistance, electrical, 258–259
revolving point, 55
right angle, 3, 5
right ascension, 152–153, 156
right-hand rule for cross product, 171–172
right triangle, 16–17
RL phase angle, 260–261

roots in scientific notation, 208–209
rounding vs. truncation, 209–210
rules for use of scientific notation,
    204–209

S

scalar, definition of, 161
scalar multiplication of vector, 161,
    165–166, 170–171
scalar product of two vectors, 162–163
scientific notation
    addition in, 204–205
    alternative, 197–198
    division in, 206–207
    exponentiation in, 208
    multiplication in, 206
    multiplication symbol in, 198
    orders of magnitude in, 200–203
    plain-text exponents in, 199
    prefix multipliers in, 202–203
    roots in, 208–209
    rules for use of, 204–209
    standard, 197
    subtraction in, 205–206
secant function, 64–65
secant of negative angle, 397
second of arc, definition of, 6
second of right ascension, definition
    of, 152–153
secondary circular functions, 62–71
sequence
    arithmetic, 326–330
    finite, definition of, 326
    geometric, 332–334
    limit of, 335–337
series
    arithmetic, 327–330

series (*Cont.*):
    for cosine function, 341–344
    for sine function, 338–341
    for tangent function, 344–348
    geometric, 332–334
    limit of, 335–337
significant figures, 215–220
similar triangles, 12
sine function
    definition of, 54–55
    series for, 338–341
sine of angular difference, 398
sine of angular sum, 397
sine of double angle, 397
sine of negative angle, 396
sine wave
    amplitude of, 57
    definition of, 56–57
    frequency of, 57
    graph of, 56–57
    shape of, 249–250
singularity, 60–61
sinusoid, 56–59, 249–250
Snell's law for refraction, 283–284
spatial coordinates, 150–159
spherical angle, 307
spherical coordinates, 155–157
spherical law of cosines, 312–313
spherical law of sines, 312
spherical mirror, 275–276
spherical polygon, 307
spherical triangle, 306–316
spiral in polar coordinates, 126–130
square units, 19
square vs. square root, 89–91
stadimetry, 227–228
standard direction angle, 52–54
standard scientific notation, 197

straight angle, 5

straight line in polar coordinates, 125, 129–130, 139–141

subscripts, use of, 196–197

subset, proper, 82

subtraction in scientific notation, 205–206

sum of two vectors, 161, 164–165, 170

superscripts, use of, 196–197

surjection, 80–81

surjective mapping, 80–81

surjective relation, 84–85

symmetry, bilateral, 105

T

tangent function
  definition of, 59–60
  series for, 344–348

tangent of double angle, 397

tangent of negative angle, 396

terahertz, definition of, 249

terrestrial coordinates, 150–151

terrestrial distance measurement, 224–229

terrestrial latitude, 150–151, 239–243, 299–305

terrestrial longitude, 150–151, 239–243, 299–305

Theorem of Pythagoras
  for cosecant and cotangent, 69–70
  for right triangles, 18–21, 35
  for secant and tangent, 67–68
  for sine and cosine, 23–24, 67–71

theta, as angle name, 4

three-space, rectangular, 153–154

total internal reflection, 281–282

transfinite number, 61

transversal to parallel lines, 226

triangle
  acute, 13–14
  adjacent sides in, 21
  definition of, 11
  equilateral, 15–16
  interior angles of, 12
  isosceles, 14–15
  naming of, 12
  obtuse, 13–14
  range of angles in, 21
  right, 16–17
  sides of, 12
  spherical, 306–316
  vertices of, 12

triangles
  congruent, 13
  similar, 12
  trigonometric, 17–24

triangulation, 229–232

trigonometric functions, definitions of, 17–18

trigonometric triangles, 17–24

truncation vs. rounding, 209–210

U

unit circle, definition of, 52–54

unit-circle paradigm, 51–73

unit hyperbola, definition of, 104–105

units squared, 19

V

variable
  dependent, 28, 83–84
  independent, 28, 83–84

variables in polar coordinates,
122–124
vector
definition of, 159–161
direction of, 159–161, 163–164, 169
magnitude of, 159–161, 163–164,
169, 173
scalar multiplication of, 161,
165–166, 170–171
vectors
cross product of two, 171–176,
214–215
dot product of two, 162–163,
166–167, 171, 213–214
scalar product of two, 162–163
sum of two, 161, 164–165, 170

vectors in Cartesian plane, 159–163
vectors in Cartesian three-space,
168–176
vectors in polar coordinates, 163–168
vectors in three dimensions, 168–176
vectors showing phase, 256–257
vernal equinox, 152
vertical-line test for function, 97
vertices of triangle, 12
virtual image, 273–274

WXYZ

wavy equals sign, 216
xyz-space, 153–154
zenith in spherical coordinates, 157